Test Bank to acompany

THE WORLD OF BIOLOGY

Fifth Edition

Solomon & Berg

Patrick Woolley
East Central College

SAUNDERS COLLEGE PUBLISHING
Harcourt Brace College Publishers
Fort Worth Philadelphia San Diego New York
Orlando Austin San Antonio Toronto
Montreal London Sydney Tokyo

Printed in the United States of America.

Wooley: Test Bank to accompany The World of Biology, Fifth Edition: Solomon & Berg.

ISBN 0-03-005954-2

567 017 987654321

Chapter 1

Introducing the World of Biology

1. Which unit of organization is indicative of all living organisms?

   A. cell
   B. tissue
   C. organ
   D. system

   Ans. A

2. Most biologists do not consider this group of organisms as living things:

   A. fungi
   B. monerans
   C. protists
   D. viruses

   Ans. D

3. Indeterminate growth refers to organisms which continue to grow indefinitely. This pattern of growth is indicative of most:

   A. animals
   B. protozoans
   C. trees
   D. bacteria

   Ans. C

4. What material is transmitted to the next generation regardless of the type of reproduction?

   A. egg
   B. sperm
   C. pollen
   D. DNA

   Ans. D

5. Growth, metabolism, response to stimuli, and adapting to environmental changes best describes:

   A. only plants
   B. only animals
   C. all living organisms
   D. all viruses

   Ans. C

6. Mechanisms that regulate the internal environment of living organisms are called:

   A. heterostatic mechanisms
   B. homeostatic mechanisms
   C. autotrophic mechanisms
   D. heterotrophic mechanisms

   Ans. B

7. Living cells are capable of:

   A. continuous motion
   B. locomotion
   C. purposeful movement
   D. self-propelled movement

   Ans. A

8. Light, temperature, need for nourishment, and chemical composition of one's surroundings represent:

   A. homeostasis
   B. adaptations
   C. stimuli
   D. metabolism

   Ans. C

9. When an organism divides in half to form two organisms the process is called:

   A. asexual reproduction
   B. sexual reproduction
   C. budding
   D. DNA transmission

   Ans. A

10. If the offspring are not a duplicate of a single parent they are a product of:

A. asexual reproduction
B. sexual reproduction
C. budding
D. evolution and adaptation

Ans. B

11. DNA is best described as:

A. the product of reproduction
B. associated primarily with asexual reproduction
C. associated primarily with sexual reproduction
D. the hereditary material

Ans. D

12. Traits that enhance an organism's ability to survive are referred to as:

A. homeostasis values
B. fitness
C. adaptations
D. evolution

Ans. C

13. A composite of all the characteristics that enable many generations of a particular organism to survive are the result of:

A. population formation
B. evolutionary processes
C. ecosystem processes
D. homeostasis

Ans. B

14. A group of individuals that belong to the same species and inhabit a given area refers to:

A. the biosphere
B. a community
C. an ecosystem
D. a population

Ans. D

15. Dogs drink water by lapping, while snakes drink by sucking. These characteristics are referred to as:

    A. adaptations
    B. evolutionary processes
    C. biological fitness
    D. genetic variations

    Ans. A

16. Most evolved characteristics require how much time to manifest themselves throughout a species?

    A. one successful generation
    B. generally two generations
    C. many generations
    D. occur within one generation as a result of mutation

    Ans. C

17. The most widely accepted mechanism that contributes to the evolutionary process is:

    A. population dynamics
    B. natural selection
    C. community selection
    D. homeostatic mechanisms

    Ans. B

18. One mechanism that selects for survival characteristics and is responsible for genetic changes within populations is:

    A. biological fitness
    B. cells
    C. environment
    D. DNA

    Ans. C

19. Algae, green plants, and some bacteria are referred to as:

    A. protists
    B. heterotrophs
    C. decomposers
    D. autotrophs

    Ans. D

20. The primary biological process that has evolved among producers is:

    A. adaptation
    B. photosynthesis
    C. decomposition
    D. heterotrophic nutrition

    Ans. B

21. Who are the two biologists that suggested plausible explanations for the mechanism of evolution?

    A. Darwin and Koch
    B. Darwin and Pasteur
    C. Darwin and Wallace
    D. Darwin and Darwin

    Ans. C

22. Natural selection is contingent upon:

    A. genetic diversity
    B. autotrophs
    C. heterotrophs
    D. poorly adapted organisms

    Ans. A

23. Most biologists recognize five Kingdoms, which are:

    A. Prokaryotea, Protista, Virusita, Plantae, and Animalia
    B. Prokaryotea, Monera, Protista, Fungi, and Animalia
    C. Prokaryotea, Protista, Fungi, Plantae, and Animalia
    D. Protista, Fungi, Plantae, Animalia, and Eukaryotea

    Ans. C

24. Which taxonomic group of organisms is incorrectly placed together in the same biological Kingdom?

    A. mushrooms and molds
    B. algae and water molds
    C. molds and yeast
    D. animals and protozoans (one-celled animals)

    Ans. D

25. Each living organism is classified by having a *Genus species* name assigned to it. This system of classification is referred to as:

    A. taxonomy system
    B. binomial system of nomenclature
    C. the standard system of nomenclature
    D. Swedish system of taxonomic nomenclature

    Ans. B

26. Which sequence of taxonomic classification is arranged in the correct hierarchy?

    A. phylum, family, order, class, genus, species
    B. phylum, order, class, family, genus, species
    C. phylum, class, order, family, genus, species
    D. phylum, class, order, family, species, genus

    Ans. C

27. What are two of the products produced in photosynthesis?

    A. carbon dioxide and water
    B. food and energy
    C. food and oxygen
    D. oxygen and water

    Ans. C

28. What are two of the reactants that react in cellular respiration?

    A. carbon dioxide and oxygen
    B. food and oxygen
    C. food and water
    D. oxygen and energy

    Ans. B

29. Two of the primary decomposers found in ecosystems are:

    A. bacteria and fungi
    B. plants and animals
    C. autotrophs and heterotrophs
    D. autotrophs and animals

    Ans. A

30. Producers and consumers are dependent upon which group of organisms for their survival?

A. autotrophs
B. heterotrophs
C. both a and b
D. decomposers

Ans. D

31. Which group of organisms is correctly classified as heterotrophs?

A. animals and algae
B. plants and fungi
C. animals and decomposers
D. algae and decomposers

Ans. C

32. When molecules are broken down and energy is released the process is referred to as:

A. cellular respiration
B. heterotrophic nutrition
C. photosynthesis
D. heterotrophic respiration

Ans. A

33. A characteristic of most living organisms is:

A. they are not limited by their resources
B. more organisms are produced than survive to reproduce
C. the environment seldom selects in modern times
D. the best-adapted do not necessarily leave more offspring

Ans. B

34. Consumers are directly dependent upon which trophic level for their nutritional needs?

A. bacteriotrophs
B. autotrophs
C. heterotrophs
D. prokaryotes

Ans. B

35. The simplest level of hierarchical organization within a living organism is:

    A. cell
    B. organelle
    C. chemical
    D. organ

    Ans. C

36. An ecosystem generally consists of:

    A. nonliving, community, consumers, producers
    B. nonliving, consumers, producers, decomposers
    C. nonliving, organisms, populations, community
    D. nonliving, plants, animals, environment

    Ans. B

37. What would the shape of the distribution of genetic traits within a species most probably resemble?

    A. circle
    B. straight line
    C. S-shaped curve
    D. bell-shaped curve

    Ans. D

38. Explain how producers, consumers, and decomposers are all interdependent upon one another.

39. Why are photosynthesis and cellular respiration considered opposites of one another?

40. Explain how long legs in a horse would be selected for through natural selection, and eventually that trait would persist throughout the species.

Chapter 2

The Process of Science

1. The word science is derived from the Latin word *scientia*, which means:

   A. Study of Life
   B. honest process
   C. to know
   D. to see wisdom

   Ans. C

2. Which of the following is not indicative of the process of science?

   A. it is creative and dynamic
   B. it changes over time
   C. it is influenced by religion
   D. it can be influenced by the personalities of scientists

   Ans. C

3. Developing hypotheses refers to:

   A. developing solutions
   B. developing educated guesses
   C. developing experiments
   D. developing theories

   Ans. B

4. The development of new technology emerges from:

   A. new theories
   B. new experiments
   C. new hypotheses
   D. new values

   Ans. A

The Process of Science

5. In the scientific process what precedes development of a hypothesis?

   A. performing experiments
   B. preliminary predictions
   C. group consensus to study new material that has not been previously researched
   D. recognize and state a problem

   Ans. D

6. Scientific methodology involves investigating the world around us:

   A. by developing new technology
   B. in a systematic way
   C. by discovering new computer programs
   D. by re-evaluating previous experiments

   Ans. B

7. Spontaneous generation is based upon:

   A. living organisms developing into nonliving material
   B. rigid scientific methodology
   C. living things arising from nonliving material
   D. math and chemistry

   Ans. C

8. Biogenesis is based upon:

   A. life arising from previously existing life
   B. life arising from nonliving material due to chemical changes
   C. religion and philosophy
   D. astrology and related disciplines

   Ans. A

9. The belief that earthworms develop from pieces of straw is indicative of

   A. biogenesis
   B. biogenesis and spontaneous generation
   C. spontaneous generation
   D. philosophy

   Ans. C

10. The Italian physician Redi demonstrated the origin of flies using:

    A. experiments that originated with spontaneous generation
    B. experiments that used scientific methods
    C. experiments that were a combination of both spontaneous generation and biogenesis
    D. data gathered from other scientists and their experiments on flies to determine the origin of flies

    Ans. D

11. Redi covered the tops of some containers with cotton mesh for the purpose of allowing:

    A. flies to enter
    B. bacteria and other microbes to enter
    C. water to enter
    D. air to enter

    Ans. D

12. Redi was able to convince most informed people that maggots:

    A. were a product of rotting meat
    B. were always present in meat, and that decaying processes released substances that activated them
    C. were products of reproduction
    D. were products of interactions with bacteria

    Ans. C

13. The final group of organisms that were involved in the spontaneous generation controversy were:

    A. plants
    B. aquatic invertebrate animals
    C. slime molds
    D. microbes

    Ans. D

14. Which individual finally resolved the controversy concerning spontaneous generation?

    A. Louis Joblot
    B. Louis Pasteur
    C. Fransesco Redi
    D. Lazaro Spallanzani

    Ans. B

The Process of Science

15. Drawing conclusions that are consistent with evidence used to support them is indicative of:

    A. logic
    B. thinking
    C. predicting
    D. experimenting

    Ans. A

16. If conclusions are inferred from generalities, the process is indicative of:

    A. deductive reasoning
    B. inductive reasoning
    C. logical conclusions
    D. incorrect logic, because scientists do not infer conclusions from generalities

    Ans. A

17. If we accept the evolutionary process as a mechanism for the development of specific traits, then we can predict that the long necks of giraffes evolved from ancestors with shorter necks, which is indicative of:

    A. deductive reasoning
    B. inductive reasoning
    C. logical induction
    D. principles of outcome

    Ans. A

18. When a general principle is discovered from data gathered from a specific example, this is indicative of:

    A. deductive reasoning
    B. inductive reasoning
    C. principles of outcome
    D. inclusive reasoning

    Ans. B

19. When it was demonstrated that there is a carrying capacity (optimal number) for largemouth bass in a pond, it could be concluded that there is a carrying capacity for other fish. This is an example of:

    A. deductive reasoning
    B. inductive reasoning
    C. principles of outcome
    D. inclusive reasoning

    Ans. B

20. Scientific *method* involves:

    A. a constantly changing set of procedures
    B. the most efficient procedure
    C. a set of ordered steps
    D. a set of steps that must be adapted to the individual scientist and the type of research

    Ans. C

21. Chance and luck are often involved recognizing a problem, but the prepared scientist may have difficulty

    A. working with other scientists
    B. getting involved in difficult research
    C. not being critical in looking at a subject
    D. viewing a subject area in a new light

    Ans. C

22. Much of Alexander Fleming's discovery of penicillin was:

    A. a good example of recognizing a problem and discovering an organism to solve the problem
    B. stating a hypothesis, and testing several organisms as potential antibodies
    C. partially luck, but he had enough knowledge and insight to recognize it had possiblities
    D. realizing that the discovery of penicillin had previously been overlooked

    Ans. C

23. The *Penicillium* that Fleming discovered was:

    A. a good example of a well-planned experiment
    B. a product of a contaminated culture
    C. one of many experiments that was tried as an antibiotic to suppress *Staphylococcus*
    D. a good example that scientific methodology was disregarded

    Ans. B

24. The first step in scientific methodology is:

    A. stating the hypothesis
    B. stating a tentative explanation
    C. preparing tentative experiments to test the hypothesis
    D. recognizing a problem

    Ans. D

The Process of Science

25. In the early stages of an investigation, a scientist frequently

    A. states many possible explanations
    B. states a single explanation
    C. restricts the investigation to a single experiment
    D. recognizes that three hypotheses are usually the most feasible

    Ans. A

26. A good hypothesis

    A. is always true
    B. must always be stated positively
    C. is falsifiable or can be proven false
    D. is always stated in a neutral manner so that it will eventually be proven to be true or false

    Ans. C

27. A good hypothesis is:

    A. supportive of recognized explanations
    B. capable of being tested and generating predictions
    C. an expansion of previous research
    D. an outgrowth of tentative research

    Ans. B

28. A prediction should be capable of:

    A. being proven as true
    B. generating positive conclusions
    C. being tested by an experiment
    D. providing positive and negative data

    Ans. C

29. When a large number of observations and experiments support a hypothesis it is called:

    A. a conclusion
    B. a principle
    C. a law
    D. a theory

    Ans. D

30. A theory that has yielded true predictions over a long period of time and has become universally accepted is referred to as:

    A. a principle
    B. a law
    C. a general theory
    D. a general conclusion

    Ans. A

31. When a principle is considered to be of great basic importance it is called:

    A. a conclusive principle
    B. a law
    C. a scientific rule
    D. an original principle

    Ans. B

32. One very important aspect of science is:

    A. truthfulness and honesty
    B. reduce time and effort
    C. minimize cost
    D. change all errors and bias

    Ans. A

33. Which process does not add new knowledge, but makes relationships among data apparent?

    A. hypothesis
    B. prediction
    C. deduction
    D. induction

    Ans. C

34. The process that produces new knowledge and generalizations is:

    A. hypothesis
    B. predictions
    C. deductions
    D. induction

    Ans. D

The Process of Science

35. Unfalsifiable hypotheses are:

    A. the foundation of scientific laws
    B. eventually made into theories
    C. impossible to disprove
    D. judged to be true

    Ans. C

36. The discovery of giant bacteria is a good example of:

    A. the role that luck plays in discoveries
    B. the role of the scientific process
    C. the role that random chance has in the logical process
    D. scientific theory being enforced

    Ans. B

37. Contrast and explain spontaneous generation and biogenesis.

38. Contrast and explain deduction and induction.

39. Explain the six steps involved in using the scientific method.

40. Differentiate between the concepts of hypothesis, theory, and principle.

Chapter 3

The Chemistry of Life: Atoms, Molecules, and Reactions

1. Instead of writing out the name of each element, a system of abbreviations has been adopted, which are called:

   A. chemical abbreviations
   B. atomic abbreviations
   C. chemical symbols
   D. chemical nomenclature

   Ans. C

2. The derivations for the abbreviations for the elements are

   A. Latin or Greek
   B. Latin or English
   C. Latin or German
   D. Greek or German

   Ans. B

3. The smallest subdivision of an element that retains its characteristic chemical properties is called a(n):

   A. atomic element
   B. atomic subparticle
   C. subatomic particle
   D. atom

   Ans. D

4. H_2O is an example of:

   A. molecule
   B. chemical ratio
   C. chemical compound
   D. elemental molecule

   Ans. C

5. O_2 is an example of:

   A. chemical equation
   B. chemical formula
   C. structural formula
   D. chemical compound

   Ans. B

6. O_3 is an example of:

   A. molecule
   B. chemical compound
   C. atomic compound
   D. structural product

   Ans. A

7. Protons and neutrons compose the:

   A. atomic shells
   B. atomic orbitals
   C. atomic configuration
   D. atomic nucleus

   Ans. D

8. Atomic physicists have discovered many subatomic particles, but only a few are considered important biologically. The number that are important are:

   A. two
   B. three
   C. four
   D. six

   Ans. B

9. For an electron to move from one level to the next, the atom must absorb a packet of energy called a(n):

   A. Bohr transition unit
   B. electron ordinate
   C. quantum
   D. electron mass unit

   Ans. C

10. $_{26}Fe$ represents the:

    A. atomic mass number
    B. atomic number
    C. atomic weight
    D. atomic nuclear number

    Ans. B

11. If an amphipathic compound has a polar and a nonpolar end, then they are both

    A. acid and base
    B. anion and cation
    C. ionically and covalently bonded
    D. hydrophylic and hydrophobic

    Ans. D

12. The space where an electron is most likely to be found is called:

    A. electron energy level
    B. electron configuration
    C. electron shell
    D. electron orbital

    Ans. D

13. Spherical and dumbbell shaped describes

    A. electron energy level
    B. electron shell
    C. electron orbital
    D. atomic nucleus

    Ans. C

14. Electron orbitals that are equal distance from the atomic nucleus are placed in the same:

    A. electron configuration
    B. electron energy level
    C. atomic mass
    D. quantum orbital

    Ans. B

15. The electron shell which is closest to the nucleus:

    A. has the greatest energy
    B. has the most variable energy
    C. has the lowest energy
    D. has the greatest number of electrons

    Ans. C

16. 1/1800 describes:

    A. mass of an atomic nucleus compared to the mass of electrons
    B. mass of a proton compared to the mass of a neutron
    C. mass of an electron compared to the mass of a proton
    D. the relationship between the atomic nucleus and atomic orbitals

    Ans. C

17. In $1^{3}H$, the number three indicates:

    A. one proton and two neutrons
    B. one proton and three neutrons
    C. one proton and three electrons
    D. atomic number and the number of isotopes

    Ans. A

18. CO_2 + H_2O ---------> H_2CO_3
    (carbon dioxide + water---> carbonic acid)

    A. CO_2 and H_2O are products
    B. H_2CO_3 is a reactant
    C. CO_2 and H_2O are reactants
    D. the arrow means equal

    Ans. C

19. Which is the correct method to show nitrogen using an electron dot schematic?

    ..
    A. :N·
    ·

    ·
    B. :H·

    ·
    C. ·N·

    ..
    D. :N:
    ..

    Ans. B

20. Which is the correct method to show carbon dioxide (CO_2) using a structural formula?

A. C–O–O
B. O–C–O
C. C=O=O
D. O=C=O

Ans. D

21. Each of the lines that are used in a structural formula (e.g. water H–O–H) represents:

A. sharing of an electron
B. sharing of a pair of electrons
C. sharing of two pairs of electrons
D. sharing of an unspecified number of electrons

Ans. B

22. The electrons in the outer shell of an atom are called:

A. valence electrons
B. ionic electrons
C. covalent electrons
D. outer shell chemical bond electrons

Ans. A

23. Which element has a complete outer shell?

A. Hydrogen (H)
B. Chlorine (Cl)
C. Argon (Ar)
D. Oxygen (O)

Ans. D

24. When an atom loses electrons it is said to be:

A. oxidized
B. reduced
C. oxidation-reduction
D. hydrolyzed

Ans. A

25. An anion forms as a result of that atom becoming:

    A. oxidized
    B. reduced
    C. disassociated
    D. bonded

    Ans. B

26. Hydration describes the process in which:

    A. water is produced
    B. compounds or molecules bond to water
    C. ionic compounds disassociate in water
    D. ions are surrounded by the opposite charged end of water

    Ans. D

27. When an atom gains an electron it is said to be:

    A. oxidized
    B. ionically bonded
    C. reduced
    D. a valence atom

    Ans. C

28. In a covalently bonded compound, an atom that has weak electronegativity would have a

    A. weak positive charge
    B. strong ionic positive charge
    C. weak negative charge
    D. strong ionic negative charge

    Ans. A

29. The reason ice floats is because:

    A. its density is greater than water at 4°C
    B. its density is less than water at 4°C
    C. its density is equal to water at 4°C
    D. it is full of carbon dioxide gas

    Ans. B

30. When a compound ionizes in water to yield hydrogen ions (H^+) it forms a(n):

A. acid
B. base
C. buffer
D. solvent

Ans. A

31. Electrolytes form as a result of the disassociation of:

A. salts and sugars
B. sugars and alcohols
C. salts and acids
D. bases and alcohols

Ans. C

32. What compound is formed when a cation other than H^+ and an anion other than OH^- ionically bond?

A. acid
B. base
C. buffer
D. salt

Ans. D

33. A solution with a pH of 3 is how many times more acidic than a solution with a pH of 7?

A. 100 times
B. 1000 times
C. 10,000 times
D. 100,000 times

Ans. C

34. Cohesion refers to hydrogen bonds forming between:

A. ions and water
B. other polar molecules and water
C. water molecules
D. ions and any polar molecule

Ans. C

35. Capillary action results from the forces of:

    A. adhesion
    B. cohesion
    C. adhesion and cohesion
    D. gravity upon polar molecules

    Ans. C

36. When water dissolves other polar compounds it is acting as a(n)

    A. solute
    B. solvent
    C. solution
    D. acid

    Ans. B

37. Explain why oxidation and reduction reactions occur simultaneously.

38. Explain how ionic bonding differs from covalent bonding.

39. Explain the difference between an electron orbital and an electron energy level.

40. Differentiate between acids, bases, and salts.

Chapter 4

The Chemistry of Life: Organic Compounds

1. The main structural compounds of cells and tissues are:

   A. molecules
   B. inorganic compounds
   C. organic compounds
   D. metabolic transformers

   Ans. C

2. The primary compounds that are indicative of living organisms are:

   A. carbohydrates, lipids, steroids, proteins
   B. carbohydrates, lipids, proteins, nucleic acids
   C. sugars, fats, deoxyribonucleic acids, enzymes
   D. sugars, ribonucleic acids, deoxyribonucleic acids, fats

   Ans. B

3. When an atom of carbon becomes stable, it will form how many covalent bonds?

   A. two
   B. three
   C. four
   D. six

   Ans. C

4. A double covalent bond represents the sharing of:

   A. two electrons
   B. three electrons
   C. four electrons
   D. six electrons

   Ans. C

The Chemistry of Life: Organic Compounds

5. The three elements that most commonly bond to carbon are:

   A. hydrogen, oxygen, nitrogen
   B. hydrogen, oxygen, potassium
   C. oxygen, calcium, phosphorous
   D. oxygen, potassium, phosphorous

   Ans. A

6. How many double and single covalent bonds will one atom of carbon form?

   A. one double and three singles
   B. two doubles and one single
   C. one double and one single
   D. one double and two singles

   Ans. D

7. Fossil fuels originate from:

   A. carbohydrates and an inorganic side group
   B. carbohydrates, lipids, and an inorganic side group
   C. hydrocarbons formed from inorganic compounds
   D. any inorganic or organic compound

   Ans. C

8. Alcohols are characterized by having a functional:

   A. carboxyl group
   B. hydroxyl group
   C. methyl group
   D. nonpolar group

   Ans. B

9. When carbon forms a double covalent bond with oxygen, the bond results in:

   A. both atoms becoming nonpolar
   B. carbon becoming nonpolar and oxygen polar
   C. carbon becoming partially negatively charged and oxygen partially positively charged
   D. carbon becoming partially positively charged and oxygen partially negatively charged

   Ans. D

10. The minimal number of functional groups which an amino acid has is:

    A. one
    B. two
    C. three
    D. four

    Ans. B

11. Polymers produced by linking organic compounds together are called:

    A. monomers
    B. condensation groups
    C. hydrolysis groups
    D. macromolecules

    Ans. A

12. Water is always a reactant in what type of reaction?

    A. condensation
    B. hydrolysis
    C. dehydration synthesis
    D. organic synthesis

    Ans. B

13. Proteins and nucleic acids are examples of:

    A. micromolecules
    B. hydrolysis reactions
    C. macromolecules
    D. monomers

    Ans. C

14. A 1:2:1 ratio is indicative of

    A. carbohydrates
    B. lipids
    C. nucleic acids
    D. proteins

    Ans. A

The Chemistry of Life: Organic Compounds

15. Compounds that have the same molecular formula but have a different molecular configuration are called:

    A. monosaccharides
    B. monomers
    C. isotopes
    D. isomers

    Ans. D

16. Glucose that is in solution exists as:

    A. aldehydes
    B. ketones
    C. hydrate
    D. ring

    Ans. D

17. Most organisms use glucose in the form of:

    A. left handed
    B. right handed
    C. ketone
    D. pentose

    Ans. B

18. When a disaccharide is hydrolyzed it yields:

    A. two monosaccharides
    B. two monosaccharides and water
    C. two monosaccharides and two water
    D. two waters

    Ans. A

19. Sucrose is a product of:

    A. glucose and lactose
    B. two glucose
    C. glucose and fructose
    D. fructose and lactose

    Ans. C

20. Carbohydrates are typically stored as:

A. monosaccharide
B. disaccharides
C. cellulose
D. starch

Ans. D

21. What is the most common carbohydrate?

A. cellulose
B. starch
C. sucrose
D. maltose

Ans. A

22. Most organisms can not hydrolyze:

A. cellulose
B. disaccharides
C. monosaccharide
D. starch

Ans. A

23. The exoskeletons of insects and crustaceans are composed of:

A. cellulose
B. chitin
C. starch
D. protein

Ans. B

24. Animals store glucose within cells in the form of:

A. cellulose
B. chitin
C. glycogen
D. disaccharides

Ans. C

25. Which compound is the most water-soluble?

A. plant starch
B. cellulose
C. lipids
D. glycogen

Ans. D

26. A hydrocarbon chain with a carboxyl on one end describes:

A. monosaccharide
B. glycerol
C. fatty acid
D. lipid

Ans. C

27. Fatty acids are divided into two primary categories:

A. monosaturated and saturated fatty acids
B. unsaturated and saturated fatty acids
C. saturated and polysaturated fatty acids
D. monoacylglycerol and triglycerol

Ans. B

28. The products of lipid hydrolysis are

A. 1 glycerol and 3 fatty acids
B. 1 glycerol, 2 fatty acids, and 1 water
C. 1 glycerol, 3 fatty acids, and 3 waters
D. 3 glycerol, 3 fatty acids, and 3 waters

Ans. A

29. The reactants in the condensation (synthesis) of a disaccharide are:

A. 2 monosaccharide
B. 2 monosaccharide, and 1 water
C. 2 monosaccharide, and 2 waters
D. 2 monosaccharide, 2 waters, and an enzyme

Ans. B

30. The products produced when two amino acids bond are:

    A. protein and 2 waters
    B. polypeptide and 2 waters
    C. dipeptide
    D. dipeptide and 1 water

    Ans. C

31. Steroids that are of biological importance are:

    A. proteins, lipids, and carbohydrates
    B. cholesterol, reproductive hormones, and bile salts
    C. cholesterol, all hormones, and proteins
    D. steroid hormones, proteins, and phospholipids

    Ans. B

32. Phospholipids consist of:

    A. two hydrophilic groups at each end
    B. one hydrophilic group and one protein
    C. one hydrophilic group and two hydrophobic groups
    D. two hydrophobic groups and a carbohydrate side group

    Ans. C

33. Amino acids bond specifically by forming:

    A. ester bonds
    B. double covalent bonds between amino acids
    C. protein bonds
    D. peptide bonds

    Ans. D

34. The correct sequence for the backbone of a polypeptide chain is:

    A. N-C-C-N-C-C-
    B. N-N-C-C-N-N-C-
    C. C-N-C-N-C-N-C-
    D. N-N-C-N-N-C-

    Ans. A

The Chemistry of Life: Organic Compounds

35. The following formula (C_{3032} H_{4816} O_{872} N_{789} S_8 Fe_4) would be representative of:

    A. carbohydrate (cellulose or starch)
    B. lipid (fat or oil)
    C. protein
    D. steroid

    Ans. C

36. A nucleotide is composed of the following

    A. pentose sugar, phosphate, and 4 nitrogenous bases
    B. pentose sugar, phosphate, and 3 nitrogenous bases
    C. pentose sugar, phosphate, nitrogenous bases and protein
    D. pentose sugar, phosphate, and nitrogenous base

    Ans. D

37. Demonstrate how two glucose molecules bond together to produce maltose.

38. What is the importance of the following?

    A. monosaccharide
    B. phospholipids
    C. amino acids
    D. nucleic acids

39. Demonstrate how a dipeptide is hydrolyzed.

40. Demonstrate how a lipid is hydrolyzed.

Chapter 5

Cell Structure and Function

1. Schleiden and Schwann's contribution to biology is:

   A. expansion of Darwin's theory of evolution
   B. described eukaryotic cell and determined the functions of many organelles that are visible with the light microscope
   C. described mitosis (cell division)
   D. proposed the cell theory

   Ans. D

2. Virchow's contribution to biology is:

   A. all cells are living
   B. all cells originate from previously existing cells
   C. the nucleus contains hereditary material which is found on the chromosomes
   D. all cells are either prokaryotic or eukaryotic

   Ans. B

3. Cells are the smallest living material capable of:

   A. sustaining themselves outside an organism and reproducing independently
   B. living independently indefinitely without assistance
   C. carrying on all of life's activities
   D. mutating and evolving into a new species of multicellular organisms

   Ans. C

4. A cell would not exist if it did not have:

   A. plasma membrane to separate it from the environment
   B. cell wall for structure and support
   C. all of the organelles essential to life
   D. suborganelles that are essential to all living organisms

   Ans. A

Cell Structure and Function

5. All organisms can be classified into two groups, based on their cell's:

   A. being capable of either carrying on autotrophic reactions, or being heterotrophic
   B. having a cell wall or not having a cell wall
   C. being prokaryotic or eukaryotic
   D. capability to carry on protein synthesis

   Ans. C

6. An organism whose cells lack most membranous bound organelles is classified as:

   A. autotrophic
   B. eukaryotic
   C. heterotrophic
   D. prokaryotic

   Ans. D

7. Many enzymes and emzymatic reactions frequently are associated with:

   A. nucleus
   B. membrane surfaces
   C. cell wall
   D. all organelles found within the cytoplasm

   Ans. B

8. The reason most cells are small is due to limitations set by:

   A. capability to reproduce
   B. capability to adequately synthesize proteins
   C. capability to release energy efficiently
   D. surface area to volume ratio

   Ans. D

9. Cells have evolved specific shapes in response to:

   A. related functions
   B. related metabolic functions
   C. their location within the organism
   D. their ability to carry on biochemical reactions

   Ans. A

10. A human egg is approximately the size of:

A. small bacteria
B. grammatical period
C. half the size of a dime
D. a nickel

Ans. B

11. Who is credited with developing the earliest microscopes?

A. Schleiden
B. Hooke
C. von Leeuwenhoek
D. Pasteur

Ans. B

12. Who is the individual that first described the cells in cork?

A. Schleiden
B. Hooke
C. van Leeuwenhoek
D. Schwann

Ans. B

13. The two features of a microscope which determine how clearly you see a specimen are:

A. magnification and density of the specimen
B. magnification and light intensity
C. magnification and resolving power
D. intensity of the stain on the specimen and the stain's ability to refract light

Ans. C

14. The ratio of the size of the image to the size of the specimen is:

A. resolving power
B. refraction index
C. optical ratio index
D. magnification

Ans. D

Cell Structure and Function

15. The minimum distance between two points at which they can be distinguished as separate and distinct, as opposed to a single blurred point describes:

    A. resolving power
    B. refraction index
    C. optical ratio index
    D. magnification

    Ans. A

16. The jelly-like material outside the nucleus of a cell is called

    A. protoplasm
    B. cytoplasm
    C. plasmic inclusions
    D. organelle components

    Ans. B

17. Organelles are suspended within:

    A. cytosol
    B. inclusions
    C. nucleoplasm
    D. reticular membranes

    Ans. A

18. The nucleus is separated from the surrounding organelles within the cell by:

    A. nucleolus, cytoskeleton, and single membranous nuclear envelope
    B. endoplasmic reticulum and nuclear envelope
    C. triple membranous nuclear envelope
    D. double membranous nuclear envelope

    Ans. D

19. Ribosomes are specifically assembled by the:

    A. nucleus
    B. chromosomes
    C. nucleolus
    D. nuclear envelope

    Ans. C

20. The function of ribosomes is:

    A. protein hydrolysis
    B. protein synthesis
    C. synthesis of nucleoplasm
    D. membrane synthesis and hydrolysis

    Ans. B

21. Enucleated cells generally:

    A. are not affected because they quickly regenerate
    B. die immediately
    C. can not carry on processes essential to the cell
    D. continue to live because combinations of other organelles are able to function independently

    Ans. C

22. When a cell begins the process of mitosis its chromatin:

    A. begins synthesizing DNA and forming chromosomes
    B. begins to coil and condense and form chromosomes
    C. begins undergoing hydrolysis so that chromosomes can be synthesized from DNA
    D. combines with the nucleolus to produce chromosomes

    Ans. B

23. Each chromosome contains thousands of:

    A. proteins
    B. proteins and ribosomes
    C. undifferentiated nucleic acids
    D. genes

    Ans. D

24. A "cookbook" with many recipes would be analogous to:

    A. chromosomes
    B. genes
    C. nucleolus
    D. nucleus

    Ans. A

Cell Structure and Function

25. Genes are composed of:

    A. chromosomes
    B. DNA
    C. DNA and RNA
    D. nucleic acids

    Ans. B

26. The instructions for producing all the proteins required by a cell are chemically coded with the:

    A. ribosomes
    B. RNA
    C. DNA
    D. nucleoplasm

    Ans. C

27. The endoplasmic reticulum is continuous with:

    A. cytoplasmic skeleton and plasma membrane
    B. nuclear envelope (outer membrane) and ribosomes
    C. nuclear envelope (outer membrane) and plasma membrane
    D. nuclear pores and the cytosol

    Ans. C

28. The endoplasmic reticulum functions in:

    A. protein synthesis and a system for transporting materials within the cell
    B. synthesis of some lipids and site of cellular respiration
    C. synthesis of proteins and cell division
    D. protein synthesis and DNA synthesis

    Ans. A

29. What organelle sorts, processes, modifies, and packages proteins?

    A. smooth endoplasmic reticulum
    B. ribosomes
    C. lysosome
    D. Golgi complex

    Ans. D

30. What organelle possesses enzymes that digest some proteins, carbohydrates, fats, and nucleic acids and destroys bacteria and other foreign material that enters a cell?

A. Golgi complex
B. vacuoles
C. lysosome
D. microbodies

Ans. C

31. What organelles are found in plant seeds that possesses enzymes that convert stored fats into sugars, which are used in the germination process?

A. vacuoles
B. glyoxysomes
C. mitochondria
D. chloroplasts

Ans. B

32. Cellular respiration occurs within these organelles:

A. rough endoplasmic reticulum
B. peroxisome
C. lysosome
D. mitochondria

Ans. D

33. What organelles are associated with photosynthesis?

A. leukoplasts
B. cristae
C. chloroplasts
D. cytoskeleton

Ans. C

34. Microtubules, microfilaments, and intermediate filaments are all associated with the

A. cytoskeleton
B. cell wall
C. plasma membrane
D. endoplasmic reticulum

Ans. A

Cell Structure and Function

35. Cellulose is the primary component of:

    A. plasma membrane
    B. cell walls
    C. biological membranes
    D. lysosome

    Ans. B

36. Thylakoids are associated with:

    A. mitochondria
    B. ribosomes
    C. chloroplasts
    D. vacuoles

    Ans. C

37. Explain why cells are typically small.

38. Explain the cell theory.

39. Compare a prokaryotic cell to an eukaryotic cell.

40. Compare the functions of:
    a. mitochondria
    b. lysosome
    c. chloroplast
    d. nucleus

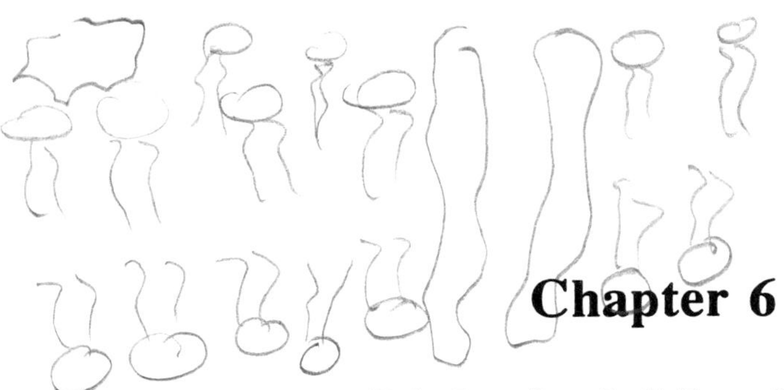

Chapter 6

Biological Membranes

1. The description of the plasma membrane as a mechanism that regulates the passage of material into and out of the cell is described as:

   A. impermeable
   B. permeable
   C. selectively permeable
   D. osmotically permeable

   Ans. C

2. Besides regulatlng permeability the cell membrane acts as a mechanism for:

   A. providing the cell with skeletal molecules that give it rigidity
   B. receiving stimuli and changes in the surrounding environment so that it can respond
   C. undergoing structural changes that provide it with the capability of changing from one type of cell to another
   D. producing proteins and lipids that assist in biochemical reactions that occur outside the cell

   Ans. B

3. Besides the plasma membrane, membranes are also associated with:

   A. nucleoli
   B. chromosomes
   C. nucleic acids
   D. organelles

   Ans. D

4. The plasma membrane could be described as a:

   A. lipid-protein mosaic
   B. protein bilayer with lipids embedded in it
   C. lipid layer and protein layer fused together
   D. lipid-protein structure in animals and cellulose structure in plants

   Ans. A

Biological Membranes

5. The lipids associated with biological membranes are:

   A. triglycerides
   B. unsaturated fats
   C. saturated fats
   D. phospholipids

   Ans. D

6. Molecules that have a hydrophobic and a hydrophilic end are called:

   A. bipolar
   B. amphipathic
   C. bivalent
   D. neutral valent

   Ans. B

7. Glycoproteins are found:

   A. on the inner surface of the plasma membrane
   B. on both the inner and outer surfaces of the plasma membrane
   C. on the outer surface of the plasma membrane
   D. in the middle of the plasma membrane

   Ans. C

8. The hydrophilic ends of a plasma membrane are located on:

   A. the inner surface or middle
   B. the outer surface
   C. the outer surface and the middle or inner surface
   D. both surfaces in a random distribution

   Ans. B

9. The center or middle of a plasma membrane is:

   A. amphipathic
   B. hydrophilic
   C. hydrophobic
   D. primarily composed of water and soluble minerals

   Ans. C

10. The lipid layers of a plasma membrane are linked together by:

    A. covalent bonds
    B. ionic bonds
    C. hydrophilic forces
    D. hydrophobic forces

    Ans. D

11. The physical state of a plasma membrane is:

    A. fluid
    B. gel
    C. solid
    D. rigid

    Ans. A

12. The lipid bilayer is impermeable to:

    A. polar and nonpolar molecules
    B. lipids
    C. water
    D. ions and polar molecules

    Ans. D

13. The proteins associated with biological membranes function as:

    A. mechanisms of support
    B. chemical transporting agents
    C. sites for producing additional proteins
    D. mechanisms for producing the entire membrane

    Ans. B

14. The part of a protein which is in contact with the lipid portion of the membrane is composed of amino acids which are:

    A. amphipathic
    B. hydrophilic
    C. hydrophobic
    D. ionic

    Ans. C

Biological Membranes

15. Peripheral proteins are usually bound to:

    A. integral proteins
    B. the hydrophilic end of a phospholipid
    C. the hydrophobic end of a phospholipid
    D. the peripheral end of microtubules

    Ans. A

16. Carbohydrates are associated with the outer surface of:

    A. phospholipids
    B. microvilli
    C. plasmodesmata
    D. glycoproteins

    Ans. D

17. The cytoplasm extensions that join adjacent plant cells are:

    A. desmosomes
    B. plasmodesmata
    C. gap junctions
    D. glycoproteins

    Ans. B

18. The surface area of plasma membranes can be increased for absorptlon of material by extensions known as:

    A. microvilli
    B. desmosomes
    C. cytoskeletal membranes
    D. extensors

    Ans. A

19. When a substance passes through a membrane the membrane is said to be:

    A. osmotic
    B. differentially permeable
    C. permeable
    D. physiologically permeable

    Ans. C

20. A membrane that allows **SOME** substances to pass through it, but does not allow other substances to pass through it is said to be

A. capable of diffusion
B. capable of osmosis
C. selectively permeable
D. hypertonic

Ans. C

21. The movement of particles from a region of higher concentration to a region of lower concentration is:

A. active transport
B. osmosis
C. diffusion
D. passive permeability

Ans. C

22. Atoms and molecules tend to diffuse in which direction along a concentration gradient?

A. up
B. down
C. across
D. through

Ans. B

23. When particles become uniformly distributed they exist in a state of:

A. equality
B. quantitative distribution
C. diffusion uniformity
D. dynamic equilibrium

Ans. D

24. Common molecules that diffuse through the plasma membrane into the cytoplasm are:

A. amino acids, glycerol, and fatty acids
B. water, oxygen, and carbon dioxide
C. glucose, proteins, and lipids
D. nucleotides, acids, and bases

Ans. B

Biological Membranes

25. Carrier proteins are associated with which process?

A. facilitated diffusion
B. osmosis
C. pinocytosis
D. gradient diffusion

Ans. A

26. Passive processes which do not require an energy expenditure by the cell include:

A. pinocytosis and osmosis
B. facilitated diffusion and phagocytosis
C. pinocytosis and phagocytosis
D. simple diffusion and facilitated diffusion

Ans. D

27. Osmosis always involves which molecule?

A. oxygen
B. carbon dioxide
C. water
D. amine

Ans. C

28. A selectively permeable membrane is indicative of what process?

A. passive diffusion
B. facilitated diffusion
C. active transport
D. osmosis

Ans. D

29. Hypertonic, hypotonic, and isotonic refer to the:

A. solute concentration inside the cell
B. solute concentration outside the cell
C. water concentration inside the cell
D. dissolved solute concentration inside the cell

Ans. B

30. Cells placed in a hypertonic solution will:

A. remain unchanged
B. swell
C. shrink
D. attain equilibrium

Ans. C

31. Internal pressure produced by hypotonic conditions in plants results in pressure being exerted against the central vacuole. This pressure is called:

A. turgor pressure
B. diffusion pressure
C. dynamic pressure
D. cellular pressure

Ans. A

32. In active transport materials move which direction?

A. they move in the same directions as materials in facilitated diffusion, but energy is required
B. higher concentration to lower concentration
C. lower concentration to higher concentration
D. if they are polar they move against the concentration gradient; if they are nonpolar, they move with the concentration gradient

Ans. C

33. Pumps that are involved in active transport are classified as:

A. phospholipid
B. adenosine triphosphate or ATP
C. sodium chloride carriers
D. integral proteins

Ans. D

34. Particles of food or entire cells that are ingested by cells is called

A. endocytosis
B. phagocytosis
C. pinocytosis
D. mediated cellularcytosis

Ans. C

B

Biological Membranes

35. When fluids that contain dissolved materials enter the cytoplasm as tiny vesicles the process is called:

A. endocytosis
B. phagocytosis
C. pinocytosis
D. mediated cellularcytosis

Ans. C

36. Materials commonly ejected from cells by exocytosis are:

A. mucus and hormones
B. metabolic wastes and lysosomes
C. metabolic wastes and water
D. metabolic wastes and sodium ions

Ans. A

37. Explain the differences and similarities between diffusion and osmosis.

38. Explain the differences and similarities between facilitated diffuslon and active transport.

39. Explain the consequences and reasons for the responses of cells placed in the following osmotic solutions:
A. hypertonic solutions
B. hypotonic solutions
C. isotonic solutions

40. Explain why the plasma membrane is described as a fluid mosaic.

Chapter 7

Solar Energy: The Power That Sustains the Planet

1. Kinetic energy differs from chemical energy in that:

   A. Kinetic energy refers to a moving object, while chemical energy refers to potential energy
   B. Kinetic energy depends upon atomic movements, while chemical energy refers to molecular movement
   C. Kinetic energy occurs in many forms and is converted into many forms of energy, while chemical energy is only converted into heat
   D. Kinetic energy is the active potential energy, while chemical energy is moving energy

   Ans. A

2. The first law of thermodynamics states:

   A. heat originates from the sun
   B. thermal energy is produced by any moving material
   C. energy can not be created or destroyed
   D. all energy eventually releases heat energy

   Ans. C

3. If a human eats 50 pounds of food over a one-month period of time, and gains 5 pounds over this one-month time period, what concept does it demonstrate?

   A. concept of chemical energy
   B. first law of thermodynamics
   C. second law of thermal dynamics
   D. third law of thermodynamics

   Ans. B

4. Living organisms are capable of transferring energy in which of the following?

   A. potential energy to light energy
   B. kinetic energy to potential energy
   C. chemical energy to light energy

   Ans. C

Solar Energy: The Power That Sustains the Planet

5. Some living organisms are capable of transferring energy in which of the following?

   A. potential energy to light energy
   B. spontaneous energy to light energy
   C. chemical energy to light energy
   D. light energy to chemical energy

   Ans. D

6. Plants and algae are capable of photosynthetic reactions, which include the following energy transfer:

   A. light energy to potential energy
   B. light energy to kinetic energy
   C. kinetic energy to chemical energy
   D. potential energy to kinetic energy

   Ans. A

7. The ultimate source of energy for living organisms is:

   A. water
   B. sunlight
   C. wind
   D. heat

   Ans. C

8. Oil and natural gas are examples of:

   A. fuels derived directly from the sun
   B. kinetic fuels
   C. fossil fuels
   D. renewable resources

   Ans. C

9. The process responsible for releasing energy from the breaking apart of organic molecules is called:

   A. photosynthesis
   B. transpiration
   C. cellular respiration
   D. chemical transformation

   Ans. C

10. The water cycle and wind (wind energy) receive energy from what source?

A. chemical energy
B. potential energy provided by photosynthesis
C. energy transformations caused by photosynthesis
D. solar energy

Ans. D

11. Which process is an example of endergonic?

A. manufacture or production of glucose from carbon dioxide and water
B. burning of wood or coal
C. glucose used in cellular respiration
D. glucose breaking down to provide energy for ATP

Ans. A

12. The glucose that is used to provide energy for a bird to fly is an example of what process?

A. kinetic energy converted to chemical energy
B. potential energy converted to chemical energy
C. cellular respiration
D. chemical energy converted to potential energy

Ans. C

13. The basic difference between exergonic and endergonic reactions is:

A. exergonic reactions release less energy than the products, while endergonic reactions release more energy than the products
B. exergonic reactions release energy and endergonic reactions absorb energy
C. exergonic reactions break covalent bonds, while endergonic reactions synthesize or produce products by covalent bonds
D. exergonic reactions break hydrogen bonds, while endergonic reactions form products that are ionically bonded

Ans. B

14. Exergonic and endergonic reactions are demonstrated when a cell uses chemical energy to:

A. produce glucose
B. produce ATP
C. break covalent bond
D. perform work

Ans. D

Solar Energy: The Power That Sustains the Planet

15. What is indicative of ATP?

    A. it consists of a three-carbon sugar and three phosphate groups
    B. it consists of three phosphate groups in which the second and third have very stable bonds
    C. it contains a six-carbon sugar and three phosphate groups in which the last two are unstable
    D. it contains a five-carbon sugar and a nitrogen compound called adenine

    Ans. D

16. Energy transformations in cells are accompanied by:

    A. release of heat into chemical energy
    B. release of heat
    C. release of heat and formation of a covalent bond
    D. formation of covalent bonds and release of ions

    Ans. B

17. The ability to produce a change in the state or motion of matter refers to:

    A. life
    B. cell
    C. energy
    D. organic compounds

    Ans. C

18. The formation of ATP from ADP occurs when it is fueled with what kinds of reactions?

    A. endergonic reactions
    B. exergonic reactions
    C. spontaneous reactions
    D. lipid synthesis reactions

    Ans. B

19. When heat is released as the result of an energy conversion, this illustrates:

    A. chemical energy
    B. first law of thermodynamics
    C. second law of thermodynamics
    D. free energy

    Ans. C

20. A measure of disorder or randomness is:

    A. entropy
    B. free energy
    C. exergonic reactions
    D. endergonic reactions

    Ans. A

21. When a change in molecular structure occurs, this is an example of:

    A. entropy
    B. spontaneous reaction
    C. dynamic equilibrium
    D. chemical reaction

    Ans. D

22. What type of reactions enable cells to control the release of free energy according to their needs, and to synthesize large macromolecules for continued use?

    A. nonreversible reactions
    B. reversible reactions
    C. nonequilibrium reactions
    D. reactions that involve ionic bonding

    Ans. B

23. ATP is different from ADP in that:

    A. ATP possesses less energy
    B. ATP is nonrenewable
    C. ATP is formed by a dehydration synthesis reaction
    D. ATP contains fewer unstable bonds

    Ans. C

24. ADP is formed:

    A. through the hydrolysis of AMP
    B. through the dehydration synthesis of ATP
    C. through the hydrolysis of ATP
    D. from ATP by an endergonic reaction

    Ans. C

Solar Energy: The Power That Sustains the Planet

25. When ATP is hydrolyzed:

A. this is an endergonic reaction
B. this is an exergonic reaction
C. energy and a molecule of water is released
D. ADP and 2Pi are produced

Ans. B

26. When a phosphate group is added to either AMP or ADP the process is referred to as:

A. exergonic reactions
B. endergonic hydrolysis reactions
C. cellular respiration
D. phosphorylation

Ans. D

27. The nature of most cellular enzymes are:

A. carbohydrates
B. lipids
C. nucleic acids
D. proteins

Ans. D

28. When an enzyme is involved in a catalytic reaction it:

A. lowers the activation energy of the reaction
B. raises the activation energy of the reaction
C. acts as one of the reactants
D. eventually becomes one of the products

Ans. A

29. The suffix that refers to an enzyme is:

A. ase
B. ise
C. ose
D. yme

Ans. A

30. The active site of an enzyme is:

    A. the location where the enzyme breaks down and becomes a product
    B. the part of the enzyme that is complementary to the substrate
    C. the part of the substrate that is modified by the enzyme
    D. the part of the enzyme that reacts with the substrate to form a new product

    Ans. B

31. The catalytic ability of enzymes are considered:

    A. fairly effective and efficient
    B. minimally effective and efficient
    C. very effective and efficient
    D. rarely effective and very inefficient

    Ans. C

32. Which materials could act as an enzyme co-factor?

    A. carbohydrates and hydrogen
    B. protein and oxygen
    C. protein and lead
    D. vitamin and manganese

    Ans. D

33. Which can affect the rate at which an enzyme reacts?

    A. carbohydrates and lipids
    B. proteins and lipids
    C. pH and temperature
    D. temperature and certain nucleic acids

    Ans. C

34. When the formation of a product inhibits an earlier reaction in a sequence of reactions, this is called:

    A. allosteric enzyme
    B. feedback inhibition
    C. coenzyme reaction
    D. enzyme sequencing

    Ans. B

Solar Energy: The Power That Sustains the Planet

35. When an enzyme is destroyed by a chemical agent it is referred to as

    A. an inhibitor
    B. a de-activator
    C. co-enzyme
    D. allosteric agent

    Ans. A

36. Many poisons and toxins are classified as:

    A. competitive inhibitors
    B. noncompetitive inhibitors
    C. sympatric inhibitor
    D. irreversible inhibitor

    Ans. D

37. Differentiate between potential energy and kinetic energy.

38. Explain how endergonic reactions differ from exergonic reactions, and how they are complementary.

39. Explain the induced-fit model as it pertains to enzymes.

40. Explain the relationship between ATP and exergonic and endergonic reactions.

Chapter 8

Energy-Releasing Pathways

1. The overall reaction for aerobic cellular respiration of glucose is:

   A. $C_6H_{12}O_6 + 12\ O_2 + 12\ H_2O \longrightarrow 6\ CO_2 + 12\ H_2O$ + Energy
   B. $C_6H_{12}O_6 + 6\ O_2 + 6\ H_2O \longrightarrow 6\ CO_2 + 12\ H_2O$ + Energy
   C. $C_5H_{10}O_5 + 5\ O_2 + 5\ H_2O \longrightarrow 5\ CO_2 + 10\ H_2O$ + Energy
   D. $2\ C_3H_6O_3 + 6\ O_2 + 6\ H_2O \longrightarrow 6\ CO_2 + 12\ O_2$ + Energy

   Ans. B

2. In the cellular respiration of glucose:

   A. glucose is reduced
   B. O_2 is reduced
   C. glucose is oxidized
   D. water is oxidized

   Ans. C

3. In the cellular respiration of glucose:

   A. water is reduced
   B. water is oxidized
   C. both water and glucose are oxidized
   D. both water and glucose are reduced

   Ans. A

4. When redox reactions occur:

   A. the substance that loses electrons becomes reduced
   B. the substance that gains electrons becomes oxidized
   C. the substance that gains electrons becomes positively charged
   D. the substance that gains electrons becomes reduced

   Ans. D

Energy-Releasing Pathways

5. During cellular respiration NADH:

   A. is involved in an exergonic reaction when it gains electrons from other molecules
   B. is involved in an endogonic reaction when it gives up electrons to other molecules
   C. is involved in an exergonic reaction when it gives up electrons to other molecules
   D. is reduced oxidized into ATP

   Ans. C

6. The process of breaking or splitting larger molecules into smaller ones is called:

   A. metabolism
   B. anabolism
   C. catabolism
   D. reduction

   Ans. C

7. During the digestive process in humans, carbohydrates are hydrolyzed into what form, so they may be absorbed by the blood?

   A. simple sugars such as sucrose
   B. simple sugars such as maltose
   C. simple sugars such as nicotinamide adenine dinucleotide
   D. simple sugars such as glucose

   Ans. D

8. One of the basic differences between aerobic pathways and anaerobic pathways is:

   A. aerobic occurs in terrestrial air breathing animals and anaerobic in aquatic animals and all plants
   B. aerobic requires O_2, and anaerobic does not require O_2
   C. aerobic does not require O_2 and anaerobic requires O_2
   D. aerobic is a catabolic reaction and anaerobic is not a catabolic reaction

   Ans. B

9. One of the differences between oxidation and reduction is:

   A. the molecule that loses electrons gives up energy and the one that gains electrons receives energy
   B. the molecule that loses electrons gains energy while the one that gains electrons loses energy
   C. the molecule that loses electrons gains energy as does the molecule that gains electrons, thus both gain energy
   D. oxidation is associated with cellular respiration while reduction is associated with photosynthesis

   Ans. A

10. One of the most common electron acceptors in biochemical pathways is

   A. phosphoglycerol aldehyde
   B. flavin adenine carbonate
   C. nitrogenous adenosine dicarbonate
   D. nicotinamide adenine dinucleotide

   Ans. C

11. $XH_2 + NAD^+$ -----> $X + NADH + H^+$ This equation demonstrates:

   A. transfer of energy from NAD^+ to compound X
   B. transfer of protons from NAD^+ to compound X
   C. transfer of electrons associated with hydrogen from compound X to NAD^+
   D. XH_2 is reduced and NAD^+ is oxidized

   Ans. C

12. Yeasts that are involved in fermentation belong to:

   A. Kingdom Plantae
   B. Kingdom Fungi
   C. Kingdom Monera
   D. Kingdom Protista

   Ans. B

13. How many chemical reactions, or groups of reactions, are associated with aerobic respiration?

   A. four
   B. eight
   C. twelve
   D. sixteen

   Ans. A

Energy-Releasing Pathways

14. Which of the following is attributed to glycolysis?

    A. splitting of glucose into pyruvic acid
    B. splitting of glucose into NAD^+
    C. splitting of glucose into FADH2
    D. reduction of glucose into ADP

    Ans. A

15. Acetyl coenzyme A (Acetyl Co A) is a product of:

    A. ATP and citric acid reduced from NADH
    B. degradation of pyruvic acid to a two-carbon fuel molecule that combines with coenzyme A
    C. degradation of glucose and O_2, to form a three-carbon acetyl group that oxidizes coenzyme A
    D. chemiosmosis and the degradation of PGAL

    Ans. B

16. The enzymes involved in the citric acid cycle are located in the:

    A. rough endoplasmic reticulum
    B. matrix of the mitochondria
    C. outer membranous surface of the mitochondria
    D. ribosomes that are associated with the mitochondria

    Ans. B

17. Which process produces the greatest amount of ATP per molecule oxidized glucose?

    A. alcoholic fermentation
    B. lactic acid fermentation
    C. aerobic respiration
    D. anaerobic respiration

    Ans. C

18. In fermentation, yeast cells:

    A. produce lactic acid before glycolysis
    B. produce lactic acid during anaerobic respiration
    C. produce alcohol before glycolysis
    D. produce alcohol after glycolysis

    Ans. D

19. A proton gradient is associated with:

 A. citric acid cycle
 B. glycolysis
 C. chemiosmosis
 D. formation of acetyl coenzyme A

 Ans. C

20. When muscles fatigue and become sore due to strenuous exercise, this is due in part to:

 A. the build up of citric acids from glycolysis
 B. the presence of lactic acid that was produced by fermentation by muscle cells
 C. the presence of ethyl alcohol that was produced by fermentation following glycolysis
 D. the presence of citric acid and ethyl alcohol that were produced by muscle cells

 Ans. B

21. Following phosphorylation a phosphorylated sugar molecule is produced, that splits in half to form two molecules of the three-carbon compound:

 A. PGAL
 B. ATP
 C. NADH
 D. pyruvic acid

 Ans. A

22. Which represents the correct sequence of events?

 A. glucose, ATP, pyruvic acid, PGAL
 B. pyruvic acid, ATP, PGAL, glucose
 C. glucose, ATP, PGAL, pyruvic acid
 D. glucose, PGAL, ATP, pyruvic acid

 Ans. D

23. The end product of glycolysis is:

 A. ATP
 B. glucose
 C. PGAL
 D. pyruvic acid

 Ans. D

Energy-Releasing Pathways

24. The citric acid cycle is synonymous with:

A. glycolysis
B. Kreb Cycle
C. oxaloacetic acid cycle
D. chemiosmosis cycle

Ans. B

25. Which of the following fuel molecules are oxidized in the citric acid cycle?

A. starches, fats, and proteins
B. sugars, saturated fats, and nucleic acids
C. ATP, PGAL, and amine groups
D. glucose, fatty acids, carbon chains of amino acids

Ans. D

26. For an animal to obtain energy from starch or any other polysaccharide, it must:

A. hydrolyze starch to sucrose
B. hydrolyze starch to ATP and amino acids
C. hydrolyze starch to glucose
D. hydrolyze starch, water, and carbon dioxide

Ans. C

27. In chemiosmosis some of the energy released from hydrogen electrons in the electron transport chains (reactions), is used to:

A. pump ATP across the inner membranes of the mitochondrion
B. pump ADP across the inner membranes of the mitochondrion
C. pump protons across the inner membranes of the mitochondrion
D. establish an electron gradient outside the mitochondrion

Ans. C

28. ATP synthesis in the electron transport system is referred to as:

A. ATP synthetase
B. ATP redox
C. ATP reduction
D. oxidative phosphorylation

Ans. D

29. Bacteria are capable of carrying on aerobic or anaerobic cellular respiration, yet they are missing:

    A. ribosomes
    B. mitochondria
    C. plasma membranes
    D. enzymes necessary for respiration

    Ans. B

30. What regulates the rate of aerobic respiration?

    A. amount of ADP and phosphate available
    B. amount of ATP and phosphate available
    C. amount of ribose, sugar, and phosphate available
    D. amount of water, phosphates, and enzymes available

    Ans. A

31. Fermentation is classified as:

    A. type of aerobic respiration
    B. type of anaerobic respiration
    C. type of endergonic reaction
    D. type of reaction restricted to unicellular organisms

    Ans. B

32. Ethyl alcohol is produced from pyruvic acid when:

    A. carbon dioxide is added to the pyruvic acid
    B. hydrogen from NADH produced during glycolysis is transferred to pyruvic acid
    C. carbon dioxide is split off from pyruvic acid and hydrogen from NADH produced during glycolysis is transferred to the two-carbon compound
    D. glucose and pyruvic acid combine with hydrogen that is split off water

    Ans. C

33. Lactic acid fermentation occurs when:

    A. fish spoil and seeds germinate
    B. canned goods spoil and top soil is produced
    C. oxygen levels increase
    D. milk sours and sauerkraut is produced

    Ans. D

Energy-Releasing Pathways

34. Bacteria and fungi that are capable of aerobic respiration, but are also capable of anaerobic respiration when there is an oxygen deficiency are called:

    A. aerobes-anaerobes
    B. facultative anaerobes
    C. obligative anaerobes
    D. strict anaerobes

    Ans. B

35. Which of the following processes is the most efficient, concerning its ability to produce the most ATP per molecule of glucose?

    A. aerobic respiration
    B. anaerobic respiration
    C. lactic acid fermentation
    D. alcohol fermentation

    Ans. A

36. The electrons that are most commonly associated with redox reactions are associated with:

    A. carbon
    B. phosphorous
    C. hydrogen
    D. nitrogen

    Ans. C

37. What role do mltochondria have in the overall role of cells?

38. What is the role of oxygen in aerobic cellular respiration?

39. What is the role of a proton gradient in chemiosmosis?

40. Explain how aerobic respiration differs from anaerobic respiration.

Chapter 9

Capturing Energy: Photosynthesis

1. Organisms classified as producers are able to:

    A. produce sugars from smaller organic compounds
    B. produce complex organic molecules from simple inorganic substances
    C. produce complex organic molecules from water and molecular oxygen
    D. produce complex organic molecules and carbon dioxide from water and light

    Ans. B

2. The few organisms that are not dependent upon photosynthesis are:

    A. yeasts
    B. protozoans
    C. sponges
    D. bacteria

    Ans. D

3. Light is a small portion of a vast continuous spectrum of radiation called the:

    A. energy spectrum
    B. solar spectrum
    C. electromagnetic spectrum
    D. electron spectrum

    Ans. C

4. The radiation form with the shortest wavelength is:

    A. gamma rays
    B. x-rays
    C. ultraviolet light rays
    D. infrared rays

    Ans. A

Capturing Energy: Photosynthesis

5. The shortest and longest wavelengths in the visible light spectrum are:

   A. blue and red light
   B. green and red light
   C. violet and red light
   D. blue and orange light

   Ans. C

6. Energy packets or particles of light are:

   A. electrons
   B. photons
   C. atoms
   D. raytons

   Ans. B

7. The lowest energy state that an electron possesses is called:

   A. neutral state
   B. electron energy level
   C. hypoenergy level
   D. ground state

   Ans. D

8. Excited electrons refer to electrons that:

   A. are raised to a higher energy level than their electron energy level
   B. are raised to a higher energy level than their hypoenergy state
   C. are raised to a higher energy level than their ground state
   D. have become covalently bonded

   Ans. C

9. A substance that absorbs light is called:

   A. spectrum absorber
   B. photoacceptor
   C. photoredox reactor
   D. pigment

   Ans. D

10. Which colors are absorbed by chlorophyll?

A. violet, blue, and red
B. violet, green, and red
C. blue, yellow, and red
D. violet, blue, and orange

Ans. A

11. Chlorophyll and other photosynthetic substances are located within the membranes of:

A. chloroplasts
B. thylakoids
C. stroma
D. smooth endoplasmic reticulum

Ans. B

12. If a scientist traces the radioactive isotopes of H and O, in which products of photosynthesis will these isotopes occur?

A. H in water and O in glucose
B. H and O will both occur in glucose
C. H in glucose and water, and O in O_2
D. H in glucose and O in water

Ans. C

13. Photosynthesis occurs in two basic sets of reactions:

A. light-dependent and light-independent reactions
B. light-dependent and light-redox reactions
C. light-dependent and photophosphorylation reactions
D. photorespiration and photophosphorylation reactions

Ans. A

14. An equation that summarizes photosynthesis is:

A. $6\ CO_2 + 6\ H_2O \xrightarrow{\text{light and chlorophyll}} C_6H_{12}O_6 + 6\ O_2$
B. $6\ CO_2 + 24\ H_2 \xrightarrow{\text{light and chlorophyll}} C_6H_{12}O_6 + 12\ H_2O + 12\ O_2$
C. $12\ CO_2 + 12\ H_2O \xrightarrow{\text{light and chlorophyll}} 2\ C_6H_{12}O_6 + 6\ H_2O + 6\ H_2O$
D. $6\ CO_2 + 12\ H_2O \xrightarrow{\text{light and chlorophyll}} C_6H_{12}O_6 + 6\ O_2 + 6\ H_2O$

Ans. D

Capturing Energy: Photosynthesis

15. When chlorophyll absorbs light energy, some of the energized electrons are transformed to chemical energy and used to:

    A. produce glucose and O_2
    B. phosphorylate Pi to AMP
    C. phosphorylate ADP to form ATP
    D. phosphorylate ADP to form PGAL

    Ans. C

16. Some of the light energy absorbed by chlorophyll is used to split:

    A. CO_2
    B. H_2O
    C. NADPH
    D. ATP

    Ans. B

17. O_2 that is produced in photosynthesis originates from:

    A. CO_2
    B. glucose
    C. chlorophyll
    D. H_2O

    Ans. D

18. What happens to the O_2 that is produced in photosynthesis?

    A. aerobic respiration in the plant and released into the atmosphere
    B. aerobic respiration in the plant and combines with carbon to form CO_2
    C. produces NADPH and released into the atmosphere
    D. reduced to form H_2O and oxidized to form glucose

    Ans. A

19. Hydrogen from water reduces:

    A. ATP
    B. ADP
    C. $NADP^+$
    D. CO_2

    Ans. C

20. Products of the light-dependent reactions of photosynthesis are

    A. ADP, $NADP^+$, O_2
    B. ATP, $NADP^+$, O_2
    C. ATP, NADPH, O_2
    D. glucose, ADP, and CO_2

    Ans. C

21. The reaction of the Calvin Cycle occur in the:

    A. thylakoid membranes
    B. stroma
    C. cytoplasm
    D. inner membranous portion of the plasma membrane

    Ans. B

22. Products of the light-independent reactions of photosynthesis are:

    A. glucose, ADP, $NADP^+$
    B. glucose, ATP, CO_2
    C. glucose, ADP, O_2
    D. NADPH, ATP, O_2

    Ans. A

23. During the Calvin Cycle the following occurs:

    A. CO_2 and $NADP^+$ are produced
    B. CO_2 and glucose are produced
    C. O_2 and glucose are produced
    D. carbon fixation and the reduction of carbon

    Ans. D

24. Photosystems I and II differ from one another in the following way

    A. photosystem I is associated with the stroma and photosystem II is not
    B. photosystem I is associated with the thylakoid membranes and photosystem II with the stroma
    C. they utilize slightly different wavelengths of light
    D. photosystem I is associated with C_3 pathways while photosystem II is associated with C_3 and C_4 pathways

    Ans. C

Capturing Energy: Photosynthesis

25. The reaction centers of P700 or P680 pigment molecules are able to:

    A. oxidize electron acceptors
    B. give up energized electrons to an electron acceptor
    C. oxidize NADPH and reduce an electron acceptor
    D. reduce NADPH and oxidize an electron acceptor

    Ans. B

26. ATP is produced by:

    A. light-dependent reaction
    B. light-independent reaction
    C. cyclic and noncyclic photophosphorylation
    D. light-independent reaction and chemosmosis

    Ans. C

27. Photosystem I is energized by light and loses a pair of electrons during noncyclic electron flow. What is the source of electrons that replace these electrons?

    A. ATP
    B. NADPH
    C. water
    D. photosystem II

    Ans. D

28. Chemosmosis that occurs during photosynthesis is different from chemosmosis that occurs during cellular respiration, because during photosynthesis:

    A. when $NADP^+$ is reduced it is the final electron acceptor
    B. an exchange of electrons occurs
    C. a proton gradient is established
    D. ATP synthase utilizes energy from H^+

    Ans. A

29. Photosystem II is energized by light and loses a pair of electrons during noncyclic electron flow. What is the source of electrons that replace these electrons?

    A. ATP
    B. NADPH
    C. water
    D. photosystem I

    Ans. C

30. The carbon that is used in photosynthesis originates from:

    A. carbohydrates
    B. fatty acids
    C. elemental carbon
    D. carbon dioxide

    Ans. D

31. The primary advantage that C_4 and CAM pathways have over C_3 pathways is:

    A. C_4 pathways are operable over a wider temperature gradient
    B. C_4 pathways are operable over a wider light spectrum
    C. C_4 pathways are able to conserve water and produce glucose more efficiently during hot drought conditions
    D. C_4 pathways are able to conserve CO_2 at night, while C_3 plants are unable to conserve respiratory CO_2

    Ans. C

32. The role of water in photosynthesis is:

    A. it provides electrons for electron transport and protons for reducing $NADP^3$
    B. it is reduced to form glucose and carbon dioxide
    C. it provides electrons for the reduction of C_2
    D. it bonds to two molecules of carbon dioxide to form a three-carbon sugar

    Ans. A

33. How is energy from electrons used by acceptors and protons in chemosmosis?

    A. used to pump H^+ from the thylakoid membranes into the stroma
    B. used to pump H^+ from the stroma across the thylakoid membrane into the interior of the thylakoid
    C. used to pump H^+ from the cytoplasm across the chloroplast membranes into the stroma of the chloroplast
    D. used to reduce PGA to PGAL

    Ans. B

34. Cyclic photophosphorylation involves only:

    A. photosystem I
    B. photosystem II
    C. the oxidation of CO_2
    D. the reduction of H_2O

    Ans. A

Capturing Energy: Photosynthesis

35. Water is split during:

    A. cyclic photophosphorylation
    B. noncyclic photophosphorylation
    C. light-independent reaction
    D. chemosmosis

    Ans. B

36. How many PGAL molecules are required to produce the six ribulose phosphate molecules that initiate the beginning of the Calvin Cycle?

    A. 6
    B. 9
    C. 10
    D. 12

    Ans. C

37. Explain why the light-independent reaction is dependent upon the llght-dependent reaction.

38. What is the role of water in the light-dependent reaction?

39. Explain the difference between C_3, C_4, and CAM biochemical pathways.

40. What is the difference between cyclic and noncyclic photophosphorylation?

Chapter 10

Biology Test

1. During prophase of mitosis, each chromosome consists of a pair of

   A. synaptic chromosomes
   B. nucleoli
   C. sister chromatids
   D. crossing-over chromatids

   Ans. C

2. The diploid chromosome complement of a female human somatic cell consists of how many autosomes and sex chromosomes?

   A. 22 autosomes and 1 X chromosome
   B. 46 autosomes and 2 X chromosomes
   C. 23 autosomes and 2 X chromosomes
   D. 44 autosomes and 2 X chromosomes

   Ans. D

3. What portion of the cell cycle occupies the greatest duration of time?

   A. interphase
   B. prophase
   C. metaphase
   D. anaphase

   Ans. A

4. Which of the following processes occur(s) during prophase of mitosis?

   A. duplication of chromosomes
   B. mitotic spindle formation
   C. cell growth
   D. homologous pairs of chromosomes synapse

   Ans. B

5. Which of the following processes occur(s) during interphase?

   A. replication of DNA
   B. division of cytoplasm
   C. crossing over between homologous chromosomes
   D. chromosomes become visible due to condensation

   Ans. A

6. The process in which two cells are produced by division of the cytoplasm is called

   A. telophase
   B. mitosis
   C. cytokinesis
   D. karyokinesis

   Ans. C

7. Which of the following processes occur(s) during metaphase?

   A. chromosomes line up along an equatorial plane
   B. the spindle mechanism forms between the centrioles
   C. nuclear membrane and nucleolus disappear
   D. centromeres of individual chromosomes divide

   Ans. A

8. Which of the following is indicative of plant cell division?

   A. contractile proteins begin forming a cleavage furrow between daughter cells
   B. cell plate begins to form between daughter cells
   C. centrioles synthesize the spindle fibers
   D. the centrosome synthesizes the centrioles within the nucleus

   Ans. B

9. Which of the following process occur(s) during telophase?

   A. chromosomes begin to condense
   B. nucleolus begins to disappear
   C. duplication of chromosomes begins
   D. the events of prophase are reversed

   Ans. D

10. Which of the following aspects of meiosis increases genetic diversity?

    A. DNA replication during interphase I
    B. pairing of homologous chromosomes
    C. crossing over of homologous chromosomes
    D. formation of tetrads

    Ans. C

11. The genes located on a particular chromosome are said to be

    A. linked
    B. synaptic pairs
    C. sister associates
    D. homologous karyotypes

    Ans. A

12. A haploid number of chromosomes is normally associated with

    A. tetrad formation
    B. somatic nuclei
    C. gametes
    D. autosomes

    Ans. C

13. Which of the following processes occur(s) during anaphase I of meiosis?

    A. homologous chromosomes separate
    B. centromeres divide and chromatids migrate to opposite poles
    C. sister centromeres split and sister chromatids migrate to opposite poles
    D. synapsis of homologous chromosomes

    Ans. A

14. Which of the following is indicative of mitosis?

    A. four daughter cells are formed
    B. homologous chromosomes remain independent or autonomous
    C. tetrads are formed which then separate into dyads
    D. cell divides twice to form haploid cells

    Ans. B

15. The chromosomes duplicate in a cell that undergoes meiosis during which stage?

    A. interphase I
    B. prophase I
    C. interphase II
    D. prophase II

    Ans. A

16. Nonsex chromosomes are called:

    A. nondisjunction chromosomes
    B. sygomatic chromosomes
    C. autosomes
    D. aneuploidies

    Ans. C

17. Synapsis of chromosomes and tetrad formation occurs during which stage of meiosis?

    A. interphase I
    B. prophase I
    C. interphase II
    D. prophase II

    Ans. B

18. Chromosome abnormalities involving one or more extra or missing chromosomes are called

    A. autosomes
    B. Down's syndrome
    C. Turner's syndrome
    D. aneuploidies

    Ans. D

19. If a chromosome is missing from a gamete, this condition has usually resulted from

    A. crossing over
    B. genetic recombination
    C. nondisjunction
    D. colchicine usage

    Ans. C

20. What procedure is used in prenatal diagnosis of genetic and chromosomal disease?

    A. analysis of anaphase I of meiosis
    B. analysis of anaphase II of meiosis
    C. analysis of telophase of mitosis
    D. amniocentesis

    Ans. D

21. If a zygote contains 14 chromosomes, how many chromosomes did the sperm contain?

    A. 28
    B. 14
    C. 7
    D. 4

    Ans. C

22. The growth phase of the cell cycle includes

    A. G_1, S, and G_2
    B. G_2, prophase, metaphase, and anaphase
    C. prophase, metaphase, anaphase, and telophase
    D. G_1 and G_2

    Ans. A

23. Free nucleotides and enzymes used in DNA replication are primarily produced during

    A. G_1 of interphase
    B. S of interphase
    C. G_2 of interphase
    D. G_0 of interphase

    Ans. A

24. If you examined 1000 root tip cells of whitefish blastulas, you would find most of the cells in:

    A. prophase
    B. metaphase
    C. telophase
    D. interphase

    Ans. D

25. In a human, the diploid number of chromosomes is 46. How many chromatids are present in a cell in late anaphase of mitosis?

    A. 12
    B. 23
    C. 46
    D. 92

    Ans. D

26. A single human sperm normally contains a complement of which sex chromosomes?

    A. XY
    B. X and Y
    C. YY
    D. XX

    Ans. B

27. The haploid number of chromosomes is first established during which phase of meiosis?

    A. anaphase I
    B. telophase I
    C. interphase II
    D. when gametes or spores are produced

    Ans. B

28. A pair of chromosomes moving to opposite poles in diploid cells is indicative of:

    A. anaphase
    B. anaphase I
    C. anaphase II
    D. telophase II

    Ans. B

29. A single pair of homologous chromosomes found in each cell that is undergoing cytokinesis is indicative of:

    A. prophase II
    B. telophase
    C. telophase II
    D. telophase I

    Ans. D

30. A single set of diploid chromosomes found in each cell that is undergoing cytokinesis is indicative of:

A. telophase
B. telophase I
C. telophase II
D. interphase

Ans. A

31. Homologous chromosomes lined up in pairs along the equatorial plate in a diploid cell is indicative of:

A. anaphase
B. metaphase
C. metaphase I
D. metaphase II

Ans. C

32. Centromeres dividing and chromosomes moving towards opposite poles in a diploid cell are indicative of:

A. anaphase
B. anaphase I
C. anaphase II
D. metaphase I

Ans. A

33. The mosquito (*Culex pipieus*) has a haploid number of 3. How many centromeres are present at anaphase I?

A. 3
B. 6
C. 12
D. 18

Ans. B

34. The mosquito (*Culex pipieus*) has a haploid number of 3. How many chromatids are present at metaphase I?

A. 3
B. 6
C. 12
D. 18

Ans. C

35. The garden pea (*Pisum satium*) has a diploid number of 14. How many chromatids are present G_2?

    A. 7
    B. 14
    C. 28
    D. 56

    Ans. C

36. The garden pea (*Pisum satium*) has a diploid number of 14. How many centromeres are present at prophase?

    A. 7
    B. 14
    C. 28
    D. 56

    Ans. B

37. If meiosis I and II are compared to mitosis, which of the meiotic processes is most like mitosis? Explain how they are similar. Also, explain how the one meiotic process is different.

38. The fruit fly *Drosophilia melanogaster* has a haploid number of 4 chromosomes. Draw a cell of this animal for each of the representative stages: prophase, anaphase, prophase I, anaphase I, and anaphase II.

39. What possible advantages does crossing over provide a species?

40. Contrast telophase in animals with that in plants.

Chapter 11

Patterns of Inheritance

1. Who is the individual that is responsible for developing the foundation of genetics?

   A. Crick
   B. Wallace
   C. Koch
   D. Mendel

   Ans. D

2. Prior to the initial work that developed the foundation of genetics, it was thought that inheritance was linked to:

   A. the reproductive cells
   B. the entire body
   C. supernatural events
   D. the weather and the climate

   Ans. B

3. If an organism has two identical alleles for a particular trait, then for that trait the organism is:

   A. heterozygous
   B. homologous
   C. homozygous
   D. recessive

   Ans. C

4. If a mouse with black hair color mates with another mouse with black hair color, and the offspring in the F_1 generation have either black coats or white coats in a ratio of 3 black:1 white, the parents must be:

   A. homozygous with black dominant to white
   B. homozygous with white dominant to black
   C. heterozygous with black dominant to white
   D. heterozygous with white dominant to black

   Ans. C

Patterns of Inheritance

5. In spiders, the allele for hairy abdomen is dominant to the allele for hairless abdomen. If two hairy individuals are crossed, their offspring would be expected to produce abdomens in the ratio of:

   A. 3 hairy abdomen : 1 hairless abdomen
   B. 2 hairy abdomen : 1 hairless abdomen
   C. 4 hairy abdomen : 1 hairless abdomen
   D. 1 hairy abdomen : 2 intermediate hairy abdomen : 1 hairless abdomen

   Ans. A

6. The genetic makeup of an individual organism constitutes its:

   A. karyotype
   B. phenotype
   C. genophytic constitution
   D. genotype

   Ans. D

The pea plant produces plants of two different sizes, and flowers that are two different colors: tall pea plants (T) are dominant to dwarf pea plants (t) while round seeds (R) are dominant to wrinkled seeds(r). Two plants heterozygous for both traits are mated, producing an F_1 population of 80 individuals. The questions below refer to this information on peas.

7. How many plants in the F_1 will produce tall plants with wrinkled seeds?

   A. 60
   B. 45
   C. 40
   D. 5

   Ans. B

8. What percentage of the F_1 will be heterozygous for both traits?

   A. 60 percent
   B. 50 percent
   C. 33 percent
   D. 25 percent

   Ans. D

9. What fraction of the plants that produce tall plants will be heterozygous for that trait?

A. 2/3
B. 1/3
C. 1/2
D. 3/4

Ans. A

10. What fraction of the plants will be dwarf and have wrinkled seeds?

A. 3/8
B. 1/4
C. 3/16
D. 1/16

Ans. D

11. What fraction of the plants will be homozygous for both traits?

A. 1/4
B. 1/2
C. 2/3
D. 3/4

Ans. A

12. The alternate form of a gene for a particular trait is called

A. genotype
B. phenotype
C. allele
D. archetype

Ans. C

13. In a monohybrid cross in which a homozygous dominant individual is crossed with a homozygous recessive individual the F_1 generation will be:

A. 3 dominant : 1 recessive
B. 2 dominant : 1 recessive
C. 1 dominant : 1 recessive
D. all dominant

Ans. D

14. In a monohybrid cross in which a heterozygous dominant individual is crossed with a homozygous recessive individual the F_1 generation will be:
    A. 3 dominant : 1 recessive
    B. 2 dominant : 1 recessive
    C. 1 dominant : 1 recessive
    D. all dominant

    Ans. C

15. Reproductive cells or sex cells are called:

    A. autosomes
    B. gametes
    C. autocytes
    D. heterocytes

    Ans. B

16. An individual that carries a genetic abnormality, but does not express the symptoms is most probably:

    A. heterozygous for the trait, but able to transmit it to offspring
    B. heterozygous for the trait, but unable to transmit it to offspring
    C. homozygous for the trait, but only able to transmit it to offspring of their own sex
    D. heterozygous for the trait, but only able to transmit it to offspring of their own sex

    Ans. A

17. The location of a particular gene on a chromosome is called:

    A. an allele
    B. a nucleotoid
    C. a locus
    D. a genome

    Ans. C

18. The only way the phenotype of a recessive trait will be expressed is:

    A. when it is homologous
    B. when it is homozygous
    C. when it is heterozygous
    D. when it is heterogenous

    Ans. B

19. The purpose of a test cross is to differentiate between:

A. individuals that are homozygous for a recessive allele
B. individuals that are heterozygous for a recessive allele
C. individuals that are homozygous for a dominant allele or heterozygous
D. individuals that are fertile and sterile

Ans. C

20. If a dihybrid cross produces an F_2 generation that has a 9:3:3:1 ratio, this represents:

A. genotypic ratio
B. phenotypic ratio
C. gamete ratio
D. hybrid ratio

Ans. B

21. When different traits or genes are located on the same chromosome, these genes are called:

A. sibling genes
B. homologous genes
C. homozygous genes
D. linked genes

Ans. D

22. Genes that are located on the same chromosome are not always inherited together, because during meiosis an event occurs that may separate these genes. That event is called:

A. crossing-over
B. segmentation
C. fragmentation
D. gene splicing

Ans. A

23. Construction of genetic maps of the entire human gene complex is called:

A. diagnostic human map
B. analytical human gene genetic map
C. human genome
D. gene targeting

Ans. C

Patterns of Inheritance

24. In the tomato, a homozygous hairless-stemmed individual is mated with a homozygous, very hairy-stemmed individual, but the offspring have scattered short hairs on their stems. This pattern of expression is most probably an example of:

    A. co-dominance
    B. incomplete dominance
    C. crossing-over
    D. pleiotropy

    Ans. B

25. How many different matings can be made in a population in which only one pair of alleles is considered?

    A. 3
    B. 4
    C. 6
    D. 8

    Ans. C

26. In humans, a downward-pointed frontal hairline (widow's peak) is a heritable trait. A person with widow's peak always has at least one parent with that trait, while individuals with a straight frontal hairline may occur in families in which one or both parents have widow's peak. When both parents have straight frontal hairline, all of their children have straight frontal hairline. How are these traits inherited?

    A. widow's peak is dominant and straight frontal hairline is incomplete
    B. widow's peak and straight frontal hairline are co-dominant
    C. widow's peak and straight frontal hairline are both incomplete dominant
    D. widow's peak is dominant and straight frontal hairline is recessive

    Ans. D

27. Opalescent dentine is a hereditary tooth disorder characterized by defective dentine that splits from stress due to normal biting and chewing. Teeth of affected persons range in color from amber to opalescent blue. Affected children occur only in families where one or both parents have opalescent dentine. Normal children have parents who are both normal, or have one normal and one affected, or both parents are affected. This gene for opalescent dentine is:

    A. completely dominant
    B. codominant
    C. incompletely dominant
    D. recessive

    Ans. A

28. A person heterozygous for type A blood marries a person heterozygous for type B blood. What types of blood could their children have?

    A. A or B blood types
    B. A, B, or AB blood types
    C. AB blood type
    D. A, B, AB, or O blood type

    Ans. D

29. What is the relationship between the genes for type A blood and type B blood in humans?

    A. incomplete dominance
    B. codominance
    C. complete dominance
    D. polygenes

    Ans. B

30. What is the relationship between the gene for type O blood and types A and B blood in humans?

    A. type O blood is an incomplete dominant to types A and B
    B. types A and B are codominants to type O blood
    C. they are all polygenes to one another
    D. type O blood is recessive to types A and B bloods

    Ans. D

31. Radishes may occur in three shapes: long, spherical, and ovoid. When a long radish is crossed with a spherical radish the F_1 generation are all ovoid radishes. What kind of inheritance does this represent?

    A. codominance
    B. incomplete dominance
    C. complete recessiveness
    D. multiple alleles

    Ans. B

Patterns of Inheritance

In some breeds of dogs, if an allele W is present the dogs have white coats, but if ww is present the dogs have colored coats. B produces black coats and bb produces brown coats. The questions below refer to the information on dog coat color.

32. In these breeds of dogs, coat color is expressed because one gene suppresses the expression of other genes (white suppresses colored coats). This is an example of:

    A. epistasis
    B. polygenes
    C. multiple alleles
    D. codominance

    Ans. A

33. A dog with the genotype wwBb would be what color?

    A. white
    B. black
    C. brown
    D. white with black and brown spots

    Ans. B

34. A dog with the genotype WwBb would be what color?

    A. white
    B. black
    C. brown
    D. white with black and brown spots

    Ans. A

35. The expression of human skin color is an expression of:

    A. codominance
    B. incomplete dominance
    C. complete dominance
    D. polygenes

    Ans. D

36. In humans, the genotype AABBCC is very dark in skin color, while the genotype aabbcc is very light in skin color. A couple who are AABBCC and aabbcc have a child which would be what genotype and phenotypic color?

    A. AABBCC and very dark
    B. AaBbCc and very dark
    C. AaBbCc and intermediate color
    D. AaBbCc and very light

    Ans. C

37. What is the difference between a chromosome, a gene, and an allele?

38. What is the difference between Mendel's principle of segregation and his principle of independent assortment?

39. What is the difference between incomplete dominance and codominance?

40. What is the difference between the following?
    A. homozygous and heterozygous
    B. genotype and phenotype
    C. epistasis and polygenes

Chapter 12

Human Genetics

1. What is the reason for sex-linked diseases being more common in men than in women?

   A. a man can express the heterozygous condition in a recessive allele and a woman can not.
   B. a man can express a recessive allele even when the dominant allele is present
   C. a man can express a single recessive allele while a woman must have two recessive alleles to express a characteristic
   D. a man will express incomplete dominance as a dominant characteristic while a woman will express it as incomplete dominance

   Ans. C

2. The diploid chromosome complement for a human female is:

   A. 46 autosomes + 2 X sex chromosomes
   B. 44 autosomes + 2 X sex chromosomes
   C. 44 autosomes + 1 X sex chromosome
   D. 22 autosomes + 1 X sex chromosome

   Ans. B

3. The diploid chromosome complement for a human male is:

   A. 46 autosomes + 2 Y sex chromosomes
   B. 44 autosomes + 2 Y sex chromosomes
   C. 44 autosomes + 1 X and 1 Y sex chromosome
   D. 22 autosomes + 1 X or 1 Y sex chromosome

   Ans. C

4. How do autosomes differ from sex chromosomes?

   A. autosomes occur as homologous pairs but differ in size depending upon the sex of the individual, whereas the sex chromosomes are not homologous
   B. autosomes and sex chromosomes are both homologous and are the same size, but differ in the characteristics they express
   C. autosomes and sex chromosomes are all the same, except the sex chromosomes only express characteristics related to sex
   D. autosomes occur as homologous pairs that are identical in size and shape, whereas the Y chromosome is smaller than the X chromosome

   Ans. D

Human Genetics

5. A human female ovum or egg has the following complement of chromosomes

   A. 22 autosomes + 1 X chromosome
   B. 22 autosomes + 2 X chromosomes
   C. 23 autosomes + 1 X chromosome
   D. 44 autosomes + 2 X chromosomes

   Ans. A

6. A human male sperm has the following complement of chromosomes:

   A. 22 autosomes + 1 X and 1 Y chromosome
   B. 21 autosomes + 1 X or 1 Y chromosome
   C. 22 autosomes + 1 X or 1 Y chromosome
   D. 44 sutosomes + 1 X and 1 Y chromosome

   Ans. C

7. In a human:

   A. the mother determines the sex of the children
   B. the father determines the sex of the children
   C. both the mother and the father determine the sex dependent upon the interaction of their proteins
   D. the pH of the mother's uterus determines the sex of the offspring, because pH associated with acidity favors males and pH associated with alkalinity favors females

   Ans. B

8. The Y chromosome in a human possesses the following characteristics:

   A. is smaller than the X chromosome and possess autosomal traits
   B. is the same size as the X chromosome but is significantly different in its shape, and possesses different traits
   C. is the same as the X chromosome but only associated with the testes and sperm
   D. is the smallest human chromosome and apparently contains genes related to the male gender

   Ans. D

9. Sex-linked genes are associated with

   A. X chromosomes only
   B. Y chromosomes only
   C. X chromosomes and all of the autosomal chromosomes
   D. X chromosomes and Y chromosomes

   Ans. A

10. Males are hemizygous, which refers to:

    A. possess only one allele of each X-linked gene rather than two
    B. possess the Y chromosome and an X chromosome
    C. their sex chromosomes are of different sizes
    D. they are an intermediate condition that occurs between homozygous and heterozygous

    Ans. A

11. How many sex chromosomes are found in a human reproductive cell?

    A. one
    B. two
    C. two pairs
    D. 46 or 23 pairs

    Ans. A

12. If a recessive trait associated with X-linkage is expressed in a female, the allele must be present:

    A. on at least one X chromosome
    B. on both X chromosomes
    C. in the mother's ovum, but would not be present in the father's sperm due to the presence of the Y chromosome
    D. in the father, but not in the mother since it is expressed as X-linkage

    Ans. B

13. Hemophilia is classified as:

    A. recessive autosomal allele
    B. dominant X-linked allele
    C. recessive X-linked allele
    D. codominant X-linked allele

    Ans. C

14. In a family of five children, one of three boys was missing his central incisors (teeth) while the remaining two boys and two girls were normal, as were the two parents. Each of the two girls had five children, and each had one son that was missing his central incisors, while the remainder of the girls and boys were normal. The most probable explanation was:

    A. the boys inherited a recessive allele
    B. the boys inherited a dominant allele
    C. the boys inherited an X-linked dominant allele
    D. the boys inherited an X-linked recessive allele

    Ans. D

15. Nystagmus or uncontrolled rolling of the eyeballs is an X-linked recessive trait in humans. Could this trait express itself in women?

    A. no, because X-linked recessive traits only express themselves in males
    B. no, because if it is X-linked it would have to be a dominant allele
    C. yes, if the mother expresses Nystagmus, because the father could be homozygous normal and the effect is offset by the mother
    D. yes, if both parents possess this recessive allele

    Ans. D

In humans autosomal chromosomes are represented as (A), and if a sex chromosome is missing it is represented as (0). A diploid female would be AAXX and a diploid male would be AAXY. A number in parentheses represents a specific autosomal chromosome. The following questions refer to this material.

16. A person with a chromosome constitution of AAXXY, would be classified as:

    A. having Down's syndrome
    B. having Klinefelter syndrome
    C. having Turner syndrome
    D. having Bookout syndrome

    Ans. B

17. A person with a chromosome constitution of AAXO, would be classified as:

    A. having Down's syndrome
    B. having Klinefelter syndrome
    C. having Turner syndrome
    D. having Valentine syndrome

    Ans. C

18. A person with a chromosome constitution of AA (21,21)XX or XY, would be classified as

    A. having Down's syndrome
    B. having Klinefelter syndrome
    C. having Turner syndrome
    D. having Dixon-Morris syndrome

    Ans. A

19. A person with a chromosome a=constitution of AAXXY, would be classified as:

    A. female
    B. male
    C. hermaphrodite
    D. pseudohermaphrodite

    Ans. B

20. A person with a chromosome constitution of AAXO, would be classified as:

    A. female
    B. male
    C. hermaphrodite
    D. mon-X Klinefelter syndrome

    Ans. A

21. A person with a chromosome constitution of AAXYY, would be classified as:

    A. female
    B. male
    C. Klinefelter syndrome
    D. Knudsen syndrome

    Ans. B

22. Persons that express a heterozygous condition for sickle-cell anemia are

    A. resistant to heart attacks
    B. more susceptible to cancer
    C. resistant to malaria
    D. susceptible to cystic fibrosis

    Ans. C

"Bent," a dominant X-linked allele (B), in mice results in a short, crooked tail; while the recessive allele (b) results in a normal tail. A normal-tailed female is mated to a bent-tailed male. The questions below refer to this material.

23. What type of female offspring would be expected?

    A. 1/2 normal tail, 1/2 bent tail
    B. 3/4 normal tail, 1/4 bent tail
    C. all normal tail
    D. all bent tail

    Ans. D

24. What type of male offspring would be expected?

    A. 1/2 normal tail, 1/2 bent tail
    B. 1/4 normal tail, 3/4 bent tail
    C. all normal tail
    D. all bent tail

    Ans. C

25. Most of the genetic diseases that occur in humans are:

    A. X-linked dominant
    B. autosomal dominant
    C. X-linked recessive
    D. autosomal recessive

    Ans. D

26. Huntington's disease is an example of a rare:

    A. X-linked dominant
    B. autosomal dominant
    C. X-linked recessive
    D. autosomal recessive

    Ans. B

27. Cystic fibrosis is an autosomal recessive allele. If two persons that were heterozygous for cystic fibrosis decided to have children, what are their chances of having a child with cystic fibrosis?

    A. 3/4
    B. 1/2
    C. 1/4
    D. none of the children would have cystic fibrosis

    Ans. C

28. Cystic fibrosis is an autosomal recessive allele. If a woman who is heterozygous for cystic fibrosis marries a man who has cystic fibrosis, what are the chances of them having a child with cystic fibrosis?

    A. 3/4
    B. 1/2
    C. 1/4
    D. all of the children would have cystic fibrosis

    Ans. B

29. Since Huntington's disease is a disease that usually afflicts middle aged persons, is there anything that can be done to reduce its frequency?

    A. no, there is nothing that can be done to prevent this disease
    B. yes, there is drug therapy that is available
    C. yes, there are surgical procedures that are available
    D. yes, the gene can be identified before the disease manifests itself and those people may choose not to have children

    Ans. D

30. One possible cure for genetic diseases and disorders possibly lies with eventually developing cures through:

    A. genetic engineering
    B. amniocentesis
    C. chorionic villus sampling
    D. preventive medicine

    Ans. A

31. What is the name of the technique that screens cultured fetal cells for chromosomal abnormalities?

    A. genetic engineering
    B. gene splicing
    C. amniocentesis
    D. karyotyping

    Ans. C

Duchenne muscular dystrophy is a recessive X-linked allele. A woman heterozygous for Duchenne marries a man who is normal. The questions below pertain to this material.

32. What percentage of the females would be heterozygous carriers of Duchenne?

    A. 3/4
    B. 1/2
    C. 1/4
    D. all of the females are carriers

    Ans. C

33. What percentage of the females would be homozygous normal?

    A. all would be normal
    B. 1/4
    C. 1/2
    D. all would be carriers

    Ans. B

34. What percentage of the males would be afflicted?

    A. all would be normal
    B. 1/4
    C. 1/2
    D. all would be afflicted

    Ans. B

35. What percentage of the males would be normal?

    A. none of the males would be normal
    B. all would be afflicted
    C. 1/4
    D. 1/2

    Ans. C

36. What percentage of the females suffer from Duchenne?

    A. 1/4
    B. 1/2
    C. 3/4
    D. none of females will be afflicted

    Ans. D

37. Why are most single gene disorders recessive rather than dominant?

38. Compare amniocentesis to chorionic villus sampling.

39. Why is hemophilia very rare in females?

40. Why are mothers carriers of X-linked recessive traits that express themselves among their sons?

Chapter 13

DNA: The Molecular Basis of Inheritance

1. Histone proteins:

   A. are the main regulators of replication
   B. protect nuclear DNA and act as the structural basis of nucleosomes
   C. are the basis for nucleosomes in ribosomes
   D. act as enzyme DNA polymerase, for adding free nucleotides onto the "unzipped" DNA molecule

   Ans. B

2. Which of the following nitrogenous bases are classified as purines?

   A. adenine and guanine
   B. adenine and thymine
   C. thymine and cytosine
   D. cytosine and guanine

   Ans. A

3. Which of the following nitrogenous bases are classified as pyrimidines?

   A. adenine and cytosine
   B. guanine and thymine
   C. cytosine and thymine
   D. adenine and quanine

   Ans. C

4. Viruses that attack or infect bacteria are referred to as:

   A. retroviruses
   B. bacterial viruses
   C. prokaryotic viruses
   D. bacteriophages

   Ans. D

5. Who was the research team that demonstrated that DNA is the genetic material of bacteriophages?

   A. Fehr and Dixon
   B. Watson and Crick
   C. Hershey and Chase
   D. Ghesedi and Cousins

   Ans. C

6. One of the most powerful techniques developed for forensic identification of a person based on their own unique form of DNA is referred to as:

   A. DNA base isolation identification
   B. DNA fingerprinting
   C. gel electrophoresis
   D. DNA profile identification

   Ans. B

7. If one strand of DNA is found to have the sequence 5' AACGTACTGC 3', what is the sequence of nucleotides on the 3', 5', strand?

   A. 3' TTGCATGACG 5'
   B. 5' TTGCTACACG 3'
   C. 3' GCAGTACGTT 5'
   D. 5' GCAGTACGTT 3'

   Ans. A

8. An individual's unique DNA can be identified if a sample of blood or:

   A. urine is available
   B. semen is available
   C. perspiration on clothing is available
   D. saliva is available

   Ans. B

9. When identifying an individual by their DNA, it is necessary to digest the DNA using a restriction enzyme, and then analyze small fragments of DNA. The technique for separating the DNA fragments by size is called:

   A. gravitational separatlon
   B. electrostatic phoresis
   C. DNA eletrolysis
   D. gel electrophoresis

   Ans. D

10. The probability of finding another person with the same DNA profile as a suspect is roughly one in:

    A. 10,000 persons
    B. 100,000 persons
    C. 10,000,000 persons
    D. 250,000,000 persons

    Ans. D

11. The giant *Acetabularia* cell is an ideal organism for studying the role of:

    A. DNA polymerase
    B. nucleotides
    C. nucleus
    D. ribosome interaction with the nucleus

    Ans. C

When the caps of *Acetabularia mediterranea* and *A. crenulata* were removed, and then their respective stalks removed and placed on the holdfasts of the opposite species, a series of experiments were performed. The questions below pertain to this material on *Acetabularia*.

12. The caps that regenerated the first time on *A. mediterranea* stalks that had been placed on *A. crenulata* holdfasts were:

    A. the caps of *A. mediterranea*
    B. the caps of *A. crenulata*
    C. a cap that was intermediate in characteristics of both *A. mediterranea* and *A. crenulata*.
    D. a cap that was completely different from either species of *Acetabularia*

    Ans. A

13. The caps that regenerated the second time on *A. mediterranea* stalks that had been placed on *A. crenulata* holdfasts were:

    A. the caps of *A. mediterranea*
    B. the caps of *A. crenulata*
    C. a cap that was intermediate in characteristics of both *A. mediterranea* and *A. crenulata*
    D. a cap that was completely different than either species of *Acetabularia*

    Ans. B

14. It was deduced from these experiments on *Acetabularia*, that what ultimately controls the cell is the:

    A. nucleus of the stalk
    B. nucleus of the cap
    C. nucleus of the holdfast
    D. combination of DNA found in the holdfasts, stalk, and cap that interacts to ultimately control the cell

    Ans. C

15. Who was the research team that determined the structure of the DNA molecule?

    A. Griffith and Hammerling
    B. Hershey and Chase
    C. Kackley and Prugh
    D. Watson and Creek

    Ans. D

16. Any alternation in the nucleotide sequence of the DNA of a gene is referred to as:

    A. mutation
    B. mutagenic agent
    C. translocator
    D. selective agent

    Ans. A

17. Griffith found that two different strains of *Streptococcus pneumonia* had different effects on mice that were injected with these two strains of bacteria:

    A. rough strain killed the mice, while the body's immunological defenses protected against the smooth strain
    B. smooth strain killed the mice, while the body's immunological defenses protected against the rough strain
    C. both the rough and smooth strains eventually killed the mice
    D. the dosage of the injection determined how the mice were affected

    Ans. B

18. If a strain of bacteria that kills mice is killed by heat, and then injected into mice, this results in:

A. the mice die at the same rate they did when injected with live bacteria
B. the mice die, but at a slower rate than they did when they were injected with live bacteria
C. some of the mice died, but some were able to develop immunity to the dead bacteria
D. none of the mice died that were injected with the dead bacteria

Ans. D

19. If a strain of bacteria that kills mice is killed with heat and injected into mice with a live strain of bacteria, that will not kill the mice, this results in:

A. many of the mice died
B. none of the mice died
C. all of the mice died immediately
D. all of the mice developed immunity to both strains

Ans. A

20. Viruses consist of the following parts:

A. plasma membrane, nucleus, chromosomes, and DNA
B. nucleus, chromosomes, and DNA
C. polysaccharide capsule, biological membrane, and nucleic acid
D. protein coat surrounding a nucleic acid core

Ans. D

21. Phage viruses penetrate the bacterial cell's outer covering and:

A. parasitize the cytoplasm of the bacterial cell
B. enter the bacterial cell, where they live within the cytoplasm
C. take over the bacterium's metabolic machinery and pathways
D. immediately kill the bacteria

Ans. C

22. Before 1952 it was not known what compound carried the phage's genetic information. Genetic information was thought to be carried by:

A. nucleic acid or DNA
B. nucleic acid or protein coat
C. DNA or RNA
D. nucleic acids or nucleosomes

Ans. B

23. It was determined which compound carried the phage's genetic information by using two radioactive isotopes. The two radioactive isotopes that were used as tracers were:

A. ^{14}C and ^{15}N
B. ^{14}C and ^{35}S
C. ^{14}C and ^{32}P
D. ^{35}S and ^{32}P

Ans. D

24. Much of the analysis of the DNA molecule was provided by a technique called:

A. X-ray chromatography
B. X-ray diffraction
C. electron microscopy
D. X-ray electrophoresis

Ans. B

25. The three-dimensional shape of the DNA molecule is described as being

A. a spiral polypeptide shape
B. a helical shape
C. a spherical shape
D. an ovoid shape

Ans. B

26. The backbone of the DNA molecule consists of alternating units of:

A. sugar and nitrogenous bases
B. phosphate and nitrogenous bases
C. sugar and phosphate
D. sugar and sulfate

Ans. C

27. The four bases of DNA are:

A. adenine, uracil, guanine, and thymine
B. adenine, guanine, cytosine, and uracil
C. adenine, guanine, cytosine, and thymine
D. sugar, phosphate, sulfate, and adenosine

Ans. C

28. DNA is referred to as the blueprint of life, because it is the molecule whose structure preprogrammed the composition and functions of:

    A. every gamete or sex cell
    B. every animal cell
    C. every plant and animal cell
    D. every living cell

    Ans. D

29. The molecular subunits of DNA are called:

    A. nucleotides
    B. nucleic acids
    C. nucleoplasm
    D. heliotides

    Ans. A

30. A deoxyribose sugar, phosphate group, and a nitrogen-containing base compound, bond together to form a unit called:

    A. double helix
    B. nucleolus
    C. nucleotide
    D. nucleosome

    Ans. C

31. The two strands of a DNA double helix are complementary:

    A. and always identical
    B. but not identical
    C. and frequently identical
    D. and seldom pair their bases

    Ans. B

32. When DNA duplicates itself the process is known as:

    A. template duplication
    B. DNA transcription
    C. semiconservative replication
    D. DNA synthesis

    Ans. C

DNA: The Molecular Basis of Inheritance

33. When DNA duplicates itself, each of the two molecules that are formed contain:

    A. two newly synthesized strands of DNA
    B. two of the old or original strands of DNA
    C. variable amounts of newly synthesized strand of DNA, depending upon the age of the cell
    D. one of the original strands and one of a newly synthesized strand

    Ans. D

34. Once DNA has divided longitudinally or "unzipped", each DNA strand acts as a template for assembling the complementary strand. Free nucleotides are added to the template by the enzyme:

    A. DNA transpotase
    B. DNA nucleoase
    C. DNA polymerase
    D. DNA templatase

    Ans. C

35. A single chromosome consists of:

    A. hundreds to thousands of DNA molecules, which each represent separate genes
    B. an indeterminate number of DNA molecules, which each represents an indeterminate number of genes
    C. a single large DNA molecule and a single gene
    D. a single large DNA molecule that codes for hundreds or thousands of different genes

    Ans. D

36. Each histone spool of proteins consists of eight histone molecules and about 140 DNA base pairs, which form a structural unit of the chromosome called a:

    A. chromatin
    B. nucleosome
    C. nucleotide
    D. template replicator

    Ans. B

37. Explain why *Acetabularie* research was important regarding the function of the nucleus.

38. Explain how DNA forms exact copies of itself during replication.

39. What is meant by semiconservative replication?

40. Draw a DNA double helix showing the sugar-phosphate backbone and each of the nitrogen base pairs.

Chapter 14

Gene Function: RNA and Protein Synthesis

1. Sickle-cell anemia is the expression of:

   A. a mutation
   B. a pathogenic protozoan transmitted by mosquitos
   C. a deficiency of iron
   D. a malaria induced condition

   Ans. A

2. People afflicted with sickle-cell anemia produce an abnormal form of:

   A. ribosomes that synthesize proteins
   B. tiny blood vessels or capillaries
   C. the blood protein, hemoglobin
   D. blood plasma

   Ans. C

3. The units of DNA and RNA that are composed of a five-carbon sugar, phosphate, and a nitrogen base compound are referred to as:

   A. nucleosomes
   B. nucleotides
   C. nucleic polymerase
   D. nucleic acids

   Ans. B

4. Any physical or chemical agent that alters the sequence of DNA nucleotides is referred to as:

   A. a transposon
   B. a codon-anticodon
   C. a terminator
   D. a mutagenic agent

   Ans. D

Gene Function: RNA and Protein Synthesis

5. DNA does not produce proteins directly but uses an intermediary referred to as:

   A. transcriptase
   B. DNA polymerase
   C. ribosomes
   D. RNA

   Ans. D

6. The process in which DNA acts as a template for coding RNA is called:

   A. translation
   B. transcription
   C. reverse transcriptase
   D. triplet-codon transferase

   Ans. B

7. When a polypeptide is formed at the ribosomes, using information encoded in RNA, it is referred to as:

   A. translation
   B. transcription
   C. polypeptide transfer
   D. protein initiation

   Ans. A

8. How does DNA differ from RNA?

   A. DNA is a double-stranded molecule, as is RNA, both contain thymine, and both utilize the same five-carbon sugar
   B. DNA is a double-stranded molecule, whereas RNA is single-stranded, both contain uracil and thymine, and they utilize the same five-carbon sugar
   C. DNA is a double-stranded molecule, whereas RNA is single-stranded, RNA contains uracil instead of thymine, and they utilize different five-carbon sugars
   D. DNA is usually single-stranded, while RNA is usually double-stranded. RNA utilizes uracil instead of thymine, and they utilize different five-carbon sugars

   Ans. C

9. If a DNA template is 5'ATCTGTTAGA 3' then the correct sequence of RNA is:

   A. 3' TAGACAATCT 5'
   B. 5' UAGAGTTAGA 3'
   C. 3' UAGACAAUCU 5'
   D. 3' AUGUCAAUGA 5'

   Ans. C

10. If a polypeptide chain contains 152 amino acids, what is the minimum number of nucleotides in mRNA coding for this chain?

A. 76
B. 152
C. 456
D. 608

Ans. C

11. Transcription pertains to the synthesis of:

A. DNA only
B. mRNA and tRNA
C. mRNA, rRNA, and tRNA
D. mRNA only

Ans. C

12. The number of possible combinations of DNA nucleotides arranged into triplets or trios of nucleotides is

A. 12
B. 16
C. 36
D. 64

Ans. D

13. This is the compound that carries the coded instruction for protein synthesis from DNA to the ribosome. It is referred to as:

A. mRNA
B. rRNA
C. tRNA
D. pDNA

Ans. A

14. This is the compound that carries specific amino acids to the ribosome during protein assembly. It is referred to as:

A. mRNA
B. rRNA
C. tRNA
D. pDNA

Ans. C

Gene Function: RNA and Protein Synthesis

15. Without this compound protein synthesis could not occur, and the ribosome would not be functional. This compound assists in the attachment of the ribosome to the compound that carries the coded instruction for protein synthesis from DNA. It is referred to as:

    A. mRNA
    B. rRNA
    C. tRNA
    D. pDNA

    Ans. B

16. When DNA acts as a template for the synthesis of a new complementary DNA strand the process is called replication. A similar process occurs when DNA acts as a template for the synthesis of a complementary RNA strand. This process is referred to as:

    A. RNA synthesis
    B. RNA replication
    C. transduction
    D. transcription

    Ans. D

17. The process by which RNA is coded or synthesized from a DNA template is controlled by a complex of enzymes called:

    A. replicationase
    B. DNA polymerase
    C. RNA polymerase
    D. promotor

    Ans. C

18. The process by which RNA is coded or synthesized from a DNA template begins when a complex of RNA enzymes binds to a site in the DNA. A special sequence of bases in the DNA that indicates where the process begins initiates the process, which is referred to as a:

    A. promotor
    B. RNA nucleotidase
    C. RNA polymerase
    D. DNA polymerase

    Ans. A

19. Since only one side of the DNA double helix transcribes, only one side serves as a template. For transcription to occur a code must be present. This code is called a:

A. transcriptionase
B. promotor sequence
C. codon
D. anticodon

Ans. B

20. A sequence of three mRNA bases which are complementary to a triplet in DNA nucleotides is referred to as:

A. introns
B. anticodons
C. codons
D. exons

Ans. C

21. When DNA transcribes base sequences into mRNA that do not code for amino acids to make a protein, theses sequences are enzymatically snipped out of the mRNA sequences and their DNA complements are called

A. exons
B. interons
C. terminators
D. excisases

Ans. B

22. Following the removal of discarded mRNA sequences and their DNA complements that do not code for amino acids to make a protein, the coded sequences that make a protein are spliced together and used at the ribosomes. These functional sequences are called:

A. exons
B. interons
C. promotors
D. transcriptionase ribosomal initiators

Ans. A

23. Translation begins when a strand of mRNA becomes attached to a ribosome. This process in translation is called:

A. elongation
B. initiation
C. promotion
D. inclusion

Ans. B

Gene Function: RNA and Protein Synthesis

24. When the ribosome moves down the mRNA strand during translation, and amino acids are added one by one to the polypeptide chain, the process is called:

A. elongation
B. initiation
C. amino acid synthesis
D. amino acid template translation

Ans. A

25. The final step in translation is when the polypeptide chain is released. This process is referred to as:

A. polypeptide separation
B. protein exitase
C. polypeptide elongation
D. termination

Ans. D

26. What brings mRNA, tRNA with their associated amino acids, rRNA, and necessary enzymes together?

A. DNA
B. chromosomes
C. ribosomes
D. nucleus

Ans. C

27. What is the structure that ensures that tRNA molecules link up with their specific amino acids? There is one of these structures for each kind of amino acid.

A. initiator
B. enzyme
C. intron
D. exon

Ans. B

28. A tRNA tag of three nucleotide bases is called:

A. an anticodon
B. a codon
C. ribosomal polymerase
D. RNA polymerase

Ans. A

29. Each ribosome is composed of:

    A. a large spherical structure in which a groove passes through the middle for reading RNA
    B. a small and large subunit that do not join together until translation
    C. two small and one large subunit that become activated when translation is initiated
    D. one small and two large subunits that are activated by mRNA

    Ans. B

30. If a mRNA strand is AUC the tRNA strand is:

    A. ATG
    B. TAG
    C. AUC
    D. UAG

    Ans. D

31. A polysome consists of the following:

    A. a complex of many ribosomes
    B. DNA and a complex of many ribosomes
    C. mRNA and multiple ribosomes
    D. mRNA, rRNA, tRNA, and multiple ribosomes

    Ans. C

32. Which of these sequences most accurately describes the pattern involved in the synthesis of a protein:

    A. DNA---mRNA---tRNA---protein
    B. DNA---mRNA---tRNA---mRNA---protein
    C. mRNA---DNA---tRNA----protein
    D. DNA----DNA----mRNA----tRNA----protein

    Ans. A

33. Termination of protein synthesis is accomplished by special stop signals. The agent that terminates protein synthesis is classified as:

    A. an anticodon
    B. a codon
    C. reverse transcriptase
    D. restriction endonuclease

    Ans. B

Gene Function: RNA and Protein Synthesis

34. When termination of protein synthesis occurs special proteins called release factors act by:

    A. releasing the ribosome
    B. changing the configuration of tRNA, mRNA, and the anticodon
    C. detaching the polypeptide chain and mRNA from the ribosome
    D. hydrolyzing the mRNA and tRNA, thus releasing the polypeptide chain

    Ans. C

35. Each strand of mRNA may be used in protein synthesis to:

    A. make a single copy of a specific protein followed by termination
    B. make two copies of a specific protein before termination
    C. make multiple copies of several different kinds of proteins
    D. make multiple copies of a particular kind of protein

    Ans. D

36. In RNA what nitrogen base compound is indicative of RNA but NOT DNA?

    A. adenine
    B. thymine
    C. uracil
    D. guanine

    Ans. C

37. Explain the differences and similarities between replication and transcription.

38. Explain the procedures involved in initiation, elongation, and termination in protein synthesis.

39. Explain how triplets, codons, and anticodons are interrelated.

40. How are DNA and RNA similar, and how are they different?

Chapter 15

Gene Regulation

1. The origin of all the cells in a multicellular organism is from a cell referred to as a:

   A. meristematic cell
   B. gamete
   C. homozygote
   D. zygote

   Ans. D

2. The shapes and functions of the numerous cells that compose a multicellular organism are different, but they are the same in one respect. That similarity is:

   A. their potency
   B. their size
   C. their genetic composition
   D. the length of time that they live

   Ans. C

3. The ability of a single cell to develop into an entire organism is called:

   A. totipotency
   B. differentiation
   C. maturation
   D. transformation

   Ans. A

4. Gene regulation is best understood in which group of organisms?

   A. animals
   B. plants
   C. protists
   D. prokaryotes

   Ans. D

Gene Regulation

5. Information that may assist in genetic engineering and medicine, that could eventually help in developing a cure for cancer and preventing birth defects ties with studies in:

   A. gene synthesis
   B. gene regulation
   C. gene polymerase
   D. gene inhibitors

   Ans. B

6. Genes that are continually transcribed throughout the life of a cell, and produce proteins that are continuously used are called:

   A. structural genes
   B. regulator genes
   C. constitutive genes
   D. obligatory genes

   Ans. C

7. Particular enzymes or other proteins are synthesized:

   A. continually in anticipation of need
   B. only when a cell specifically requires a particular protein
   C. and stored within the cytoplasm of the cell, so that when the need for a specific enzyme or protein is required, it is available
   D. before the cell becomes functional, thus a cell possesses a complete complement of proteins and enzymes shortly after it is produced

   Ans. B

8. What organism is most commonly used in studies involving the understanding of gene regulation?

   A. *Homo sapiens*
   B. *Eurycea lucifuga*
   C. *Marmata monax*
   D. *Escherichia coli*

   Ans. D

9. Most, if not all, bacterial genes are organized into clusters called:

   A. operons
   B. DNA transcriptase
   C. DNA polymerase
   D. DNA and RNA polymerase

   Ans. A

10. Several genes that code for the synthesis of a group of enzymes that are all involved in the same function are called:

    A. constitutive genes
    B. inductive genes
    C. structural genes
    D. regulator genes

    Ans. B

11. That part of the operon where RNA polymerase binds to the DNA to begin transcription is referred to as:

    A. a structural gene
    B. a promotor
    C. an initiator
    D. a regulator

    Ans. B

12. What part of an operon determines whether structural genes will be transcribed?

    A. regulator
    B. promotor
    C. inductor
    D. operator

    Ans. D

13. The operator gene is controlled by a special part of the operon called

    A. a regulator gene
    B. a repressor gene
    C. a structural gene
    D. an inducer gene

    Ans. A

14. What binds to an operator, that prevents RNA polymerase from binding to the promotor region of the operon?

    A. regulator gene
    B. repressor gene
    C. antioperon gene
    D. depressor gene

    Ans. B

Gene Regulation

15. When RNA polymerase binds to a promotor it transcribes all of the structural genes onto one:

A. tRNA
B. rRNA
C. mRNA
D. regulator gene

Ans. C

16. The part of an operon that acts as a "control switch" that can switch transcription of the structural genes on and off by preventing RNA polymerase from binding to the promotor is:

A. regulator
B. operator
C. repressor
D. tRNA polymerase

Ans. D

17. Regulator genes, operators, promotors, and structural genes are all components of:

A. mRNA
B. rRNA
C. cluster genes
D. operons

Ans. D

18. Transcription of mRNA from DNA begins with recognition of the promotor site on DNA by:

A. RNA transcriptase
B. RNA polymerase
C. operons
D. regulator genes

Ans. B

19. Lactose operons in *E. coli* is activated by the presence of:

A. RNA polymerase
B. tRNA transcriptase
C. mRNA
D. lactose

Ans. D

20. "Lac" operon is synonymous with:

    A. lactose operon
    B. lactase operon
    C. lactase enzymes
    D. operon enzyme complex for transcription of all proteins

    Ans. A

21. A "lac operon" consists of three structural genes and:

    A. a regulator, operator, and repressor
    B. RNA polymerase, operator, and repressor
    C. a regulator, RNA polymerase, and three operons
    D. regulator, gene, operator, and promotor sites

    Ans. D

22. The structural genes that code for three enzymes used in the digestion of lactose have three distinct functions. The FIRST enzyme functions by:

    A. hydrolyzing lactose to simpler enzymes
    B. hydrolyzing the plasma membrane so that lactose can bond to the hydrolyzed site
    C. serving as an enzyme that transports lactose across the plasma membrane
    D. altering the molecular configuration of lactose

    Ans. C

23. The structural genes that code for three enzymes used in the digestion of lactose have three distinct functions. The second enzyme functions by:

    A. breaking down lactose into simpler starches
    B. breaking down lactose into simple sugars
    C. hydrolyzing lactose to simple sugars
    D. altering the configuration of lactose

    Ans. B

24. In the digestion of lactose in *E. coli*, the structural genes are:

    A. regulated
    B. unregulated
    C. absent
    D. repressed

    Ans. A

Gene Regulation

25. In the digestion of lactose in *E. coli*, the regulator gene that codes for the repressor protein is:

    A. constitutive, or the regulator is always turned on and continually makes a small amount of the repressor protein
    B. constitutive, or the repressor gene is always turned on and continually making small amounts of operon
    C. mRNA polymerase that codes for regulation of lactose enzymes
    D. one that regulates the amount of lactose that passes across the plasma membrane into the cell

    Ans. A

26. If lactose is absent from *E. coli*, a repressor protein binds to the operator site, thus blocking the:

    A. constitutive site in the process
    B. structural site in the process
    C. promotor site in the process
    D. synthesis of RNA polymerase

    Ans. C

27. If RNA polymerase cannot bind to the promotor site, it cannot begin to transcribe the

    A. operator genes
    B. regulator genes
    C. repressor genes
    D. structural genes

    Ans. D

28. When there is no lactose present in *E. coli* the structural genes that code for the enzymes needed to digest lactose are:

    A. stored in their entirety within the cell until a source of lactose becomes available
    B. deactivated, by changing the molecular configurations, and stored until a source of lactose becomes available
    C. remain active within the cell by associating with other similar substrates
    D. not even transcribed

    Ans. D

29. When lactose becomes available within *E. coli*'s environment, the following occurs:

   A. the operator genes are turned on
   B. the operon is turned on
   C. the repressor is activated
   D. the regulator gene is modified, thus becoming active

   Ans. B

30. When lactose becomes available within *E. coli*'s environment, a few molecules of lactose enter the cell and are:

   A. hydrolyzed immediately to provide a source of energy for synthesizing
   B. converted to a derivative of lactose
   C. immediately synthesized into a starch-like compound that is capable of releasing more energy
   D. converted to regulator genes and operator genes

   Ans. B

31. In the digestion of lactose in *E. coli*, what attaches to the repressor protein, thereby modifying its shape so that the repressor protein can no longer bind to the operator site?

   A. a derivative of lactose
   B. a modifier gene
   C. RNA polymerase
   D. mRNA

   Ans. A

32. In the digestion of lactose in *E. coli*, what binds to the promotor site when it is unblocked?

   A. free enzymes that hydrolyze lactose
   B. DNA polymerase
   C. RNA polymerase
   D. regulator gene

   Ans. C

33. Once the promotor site is unblocked, it can begin to

   A. produce operons
   B. produce regulators
   C. suppress the repressor
   D. transcribe structural genes

   Ans. D

Gene Regulation

34. What is it called when the presence of a substrate induces the synthesis of an enzyme?

    A. substrate operon
    B. inducible system
    C. substrate-induced synthesis
    D. enzyme-stimulated response system

    Ans. B

35. In eukaryotes, when DNA is coiled around histines to form nucleosomes it cannot:

    A. replicate
    B. grow
    C. transcribe RNA
    D. undergo mitosis or meiosis

    Ans. C

36. When some enzymes are synthesized they are in an inactive form and must be converted to an active form by the removal of a portion of the polypeptide chain. This process is referred to as:

    A. post-translational processing
    B. repressor processing
    C. enzyme conversion-activation
    D. post-synthesis activation

    Ans. A

37. Describe the operation and function of an operon.

38. How does prokaryotic gene regulation differ from eukaryotic gene regulation?

39. Differentiate between inducible, repressible, and constitutive genes.

40. What binds to the following and explain what occurs:
    a. promotor
    b. operator
    c. repressor

Chapter 16

Out With the Bad Genes, In With the Good

1. Part of the problem in treating serious genetic disorders lies in the fact that there are:

   A. no methods for detecting them
   B. no cures for them
   C. methods of detection, but since they are fatal among newborns it must be administered soon after birth
   D. too few individuals that are afflicted, thus treatment is difficult to administer

   Ans. B

2. When normal copies of a gene are introduced into some cells of the body of a person afflicted with a genetic disorder, the techniques is referred to as:

   A. gene transplant technology
   B. favorable gene expression technology
   C. gene alleviation therapy
   D. gene replacement therapy

   Ans. D

3. Children who are afflicted with adenosine deaminase (ADA), which is a severe immune deficiency disease caused by the inability of cells in the immune system to produce an enzyme can be treated by introducing the normal ADA gene into their white blood cells. Introducing the normal ADA gene is done by:

   A. using a virus to introduce the normal gene
   B. using pressure administered by a needle-less hypodermic
   C. directly injecting the normal genes into lymph nodes where the white blood cells are produced
   D. transfusing white blood cells that have been treated with heat to introduce the DNA of normal ADA genes

   Ans. A

Out With the Bad Genes, In With the Good

4. When children are treated for ADA by introducing the normal ADA gene the method is:

   A. considered a permanent procedure that has cured the deficiency
   B. considered a permanent procedure, but the patient must take antibiotic daily as a supplement
   C. considered temporary, because the patient must receive normal genes every two to three months
   D. considered temporary, because the patient must receive normal genes daily along with an antibiotic supplement

   Ans. C

5. Diseases that might be treated by introducing normal genes to replace defective genes are:

   A. malaria, cancer, and hemophilia
   B. muscular dystrophy, cystic fibrosis, and whooping cough
   C. sickle-cell anemia, hemophilia, and cystic fibrosis
   D. Parkinson's disease, AIDS, and measles

   Ans. C

6. When farmers allow only certain animals to breed, or save certain seeds for breeding purposes, the practice is referred to as:

   A. natural selection
   B. selective breeding
   C. genetic engineering
   D. selective biotechnology breeding

   Ans. B

7. Problems associated with farmers selecting certain organisms for breeding are:

   A. it requires a long time to obtain results, and they select the wrong traits
   B. they do not always achieve the results they attempt to accomplish, and their organisms lack genetic diversity
   C. organisms possess too much genetic diversity to obtain good results, and the organisms frequently are incompatible biochemically
   D. they do not always achieve the results they attempt to accomplish, and it requires a long time to obtain results

   Ans. D

8. When living organisms are used to produce products that benefit humanity it is known as:

   A. biotechnology
   B. genetic engineering
   C. recombinant DNA technology
   D. genetic replacement technology

   Ans. A

9. When the DNA of an organism is modified to produce new characteristics or traits within that organism, it is known as:

   A. biotechnology
   B. recombinant DNA technology
   C. DNA splicing
   D. genetic engineering

   Ans. D

10. When the DNA of an organism is modified to produce new characteristics or traits within an organism, it is superior to farmers or individuals selecting for certain organisms for breeding because it differs in that:

   A. the results are always predictable
   B. a much larger gene pool becomes available for use
   C. only defective genes are focused upon
   D. mistakes are eliminated and the organism is quickly improved

   Ans. B

11. Methods that are utilized or employed in genetic engineering are known as:

   A. biotechnology
   B. DNA vector analysis
   C. recombinant DNA technology
   D. manipulative DNA technology

   Ans. C

12. When new combinations of genes are produced by isolating genes from one organism and introducing them into either a similar or an unrelated organism, it is referred to as:

   A. DNA insertion
   B. recombinant DNA technology
   C. additive DNA technology
   D. gene splicing

   Ans. B

Out With the Bad Genes, In With the Good

13. Once new or foreign DNA in inserted into a single cell bacterium, it may

    A. be expressed
    B. work with the DNA present within the cell to form new gene combinations
    C. eliminate the expression of the bacterial genes
    D. produce new combinations of proteins that are expressions of the inserted DNA and the bacterial DNA

    Ans. A

14. Recombinant DNA technology is not a practice that is restricted to humans, or a laboratory procedure. Transfer of DNA occurs between:

    A. bacteria and fungi
    B. bacteria and protozoans
    C. plants and bacteria
    D. one bacterium to another by a bacterial virus

    Ans. D

15. To isolate the gene of interest, a cell's DNA must be broken up into more manageable fragments by the use of enzymes produced by:

    A. a virus
    B. a bacterium
    C. a fungus
    D. a phage virus

    Ans. B

16. The enzyme that breaks the DNA up into manageable fragments, when a researcher is attempting to isolate a particular gene, is referred to as:

    A. fragmentase enzyme
    B. ligase enzyme
    C. restriction enzyme
    D. constrictionase enzyme

    Ans. C

17. Enzymes that break DNA into manageable fragments, cut DNA molecule

    A. between the phosphoric acid and the pentose sugar
    B. between the pentose sugar and specific anticodons
    C. at specific base sequences
    D. at specific codons

    Ans. C

18. Enzymes that break DNA into fragments have evolved as defense mechanisms against:

    A. viruses
    B. bacteria
    C. humans
    D. algae and plants or photosynthetic organisms

    Ans. B

19. One obstacle or problem that had to be overcome through the evolutionary process, when enzymes evolved to break DNA into fragments was:

    A. specific anticodons must be recognized
    B. specific codons must be recognized
    C. specific enzymes must be synthesized as the need arises
    D. DNA is modified so that it is immune to its own enzymes

    Ans. D

20. If an enzymes recognizes the DNA base-pair sequence 5"-GGACTT-3" it would recognize the sequence:

    A. 3"-CCTGAA-5"
    B. 5"-CCTGAA-3"
    C. 3"-TTCAGG-5"
    D. 5"-AAGTCC-3"

    Ans. A

21. When an enzyme recognizes the base sequence of one strand and its complementary strand, it is referred to as:

    A. recombinant recognition
    B. DNA reverse recognition
    C. palindromic
    D. DNA

    Ans. C

22. That portion of a DNA segment that can pair with the single-stranded ends of other DNA molecules that were digested by the same enzyme are referred to as:

    A. restriction ends
    B. sticky ends
    C. adhesive ends
    D. splicing ends

    Ans. B

Out With the Bad Genes, In With the Good

23. The DNA base sequences that are cut are always the same for each enzyme. The DNA base sequences that will be cleaved

    A. will vary from one species of organisms to the next
    B. will vary between prokaryotic and eukaryotic cells
    C. are randomly broken depending upon the particular species and its DNA
    D. are always the same regardless of the type of DNA

    Ans. D

24. The complementary ends of DNA segments may be spliced together by a splicing enzyme known as:

    A. DNA splicase
    B. DNA ligase
    C. DNA polymerase
    D. DNA palindromicase

    Ans. B

25. Plasmids, viruses, and gene guns are all classified as:

    A. a vector or DNA carrier
    B. phages
    C. amplifiers or recombinants
    D. DNA modifiers

    Ans. A

26. Bacteria that have been treated with heat and calcium, and which take in or receive plasmids, are said to be:

    A. altered
    B. reduced
    C. transformed
    D. recombined

    Ans. C

27. To determine which bacteria have received the genetically altered plasmids, or carry the intended genetic information, they are identified by using:

    A. particular radioactive isotopes
    B. particular biological stains
    C. known DNA sequences that will bond to complementary segments
    D. particular antibiotics that they will resist

    Ans. D

28. A genetic probe is a radioactively labeled:

    A. segment of RNA or single-stranded DNA
    B. phosphoric acid and pentose sugars
    C. isotopes associated with proteins
    D. group of amino acids and tRNA anticodons

    Ans. A

29. The gene of interest will not be expressed within the new cell or transcribed within the new cell unless it is associated with a set of:

    A. specific enzymes and proteins
    B. regulatory and promoter genes
    C. operator genes and operons
    D. introns and specific amino acids

    Ans. B

30. When mRNA is isolated, and the enzyme reverse transcriptase is used to make DNA copies from mRNA, the DNA is referred to as:

    A. mRNA transcribed DNA
    B. reverse DNA transcription
    C. complementary DNA
    D. isolated reverse DNA

    Ans. C

31. The process in which millions of copies of DNA can be produced from a tiny sample or amplified from a small sample is referred to as:

    A. DNA replication-transcription
    B. DNA polyassembly
    C. DNA multiple duplication process
    D. polymerase chain reaction

    Ans. D

32. The production of human insulin by *E. coli* or an intestinal bacteria is an example of:

    A. a genetically engineered protein
    B. recombinant RNA engineering
    C. cDNA technology
    D. pharmaceutic research

    Ans. A

Out With the Bad Genes, In With the Good

33. One of the most common vectors used to introduce genes into plants is the

    A. crown gall virus
    B. crown gall bacterium
    C. crown gall hormone
    D. DNA polymerase

    Ans. B

34. After a multicellular plant has been genetically transformed it must be grown:

    A. in a greenhouse
    B. by tissue culture
    C. by hydroponics
    D. in the presence of the vector that was involved in transforming it genetically

    Ans. B

35. One of the benefits that is derived from a genetically engineered crop such as corn that is resistant to the European corn borer is:

    A. it does not harm the environment and require the application of pesticides
    B. it will simplify the food chain by reducing the presence of corn borers
    C. it will reduce the cost of plowing and thus reduce the cost of fuel
    D. it will give the corn a competitive edge against weeds, thus reducing the cost of herbicides.

    Ans. A

36. If a gene is inserted into cancer cells and causes the cancer cell to die within the cancer patient, this is an example of:

    A. biotechnology
    B. genetic engineering
    C. recombinant DNA technology
    D. gene replacement therapy

    Ans. D

37. Differentiate between the following:
    a. gene replacement therapy
    b. genetic engineering
    c. biotechnology
    d. recombinant DNA technology

38. Explain what restriction enzymes are, and how they are used in DNA research.

39. Differentiate between a eukaryotic gene and complementary DNA (cDNA).

40. Explain the methodology involved in the PCR method of amplifying DNA.

Chapter 17

Darwin and Natural Selection

1. Evolution is a process that:

   A. occurred long ago and has resulted in the present forms of life
   B. is a constant ongoing process that continues to express evolutionary changes
   C. is an ongoing process among microbial organisms, but no longer expresses itself among plants and animals
   D. is only adequately demonstrated by fossils

   Ans. B

2. Evolution is considered by most biologists to be:

   A. a process that links all fields of biology
   B. a process regarded with skepticism
   C. a process that occurred in the past but is difficult to link to the present
   D. a process that has application to the lower forms of life, but has little application to the higher forms of life

   Ans. A

3. Evolution is defined as:

   A. genetic changes that occur within the lifetime of an organism
   B. changes in the appearance of an organism
   C. changes in the environment and climatic condition of the earth
   D. genetic changes in a population of organisms that occur over time

   Ans. D

4. Over time, a population changes:

   A. as favorable traits and less favorable traits attain equilibrium
   B. in the appearance of the organisms, but does not demonstrate genetic changes
   C. as favorable traits increase in frequency and less favorable traits decrease in frequency
   D. as favorable traits increase in frequency, they eventually perfect the organism to the point where evolution is no longer applicable

   Ans. C

Darwin and Natural Selection

5. It is generally accepted that all forms of life descended from:

   A. specific life forms, such as animals evolved from animal, and plants evolved from plants
   B. either plants or animals
   C. funguslike organisms
   D. one or a few kinds of simple organisms

   Ans. D

6. When Charles Darwin used the phrase "descent with modification", it was another way of expressing a term for:

   A. evolution
   B. natural selection
   C. fitness
   D. adaptation

   Ans. A

7. Another way to describe evolution is to say that it involves changes in the frequencies of

   A. certain alleles
   B. certain proteins
   C. certain types of cells
   D. certain populations

   Ans. A

8. All the genes present within a population at a given moment in time is known as:

   A. a genome
   B. a gene pool
   C. a gene frequency
   D. a complement of genes

   Ans. B

9. Who is the individual that visualized living organisms as being imperfect but moving toward a more perfect state?

   A. Charles Darwin
   B. Alfred Wallace
   C. Aristotle
   D. Leonardo da Vinci

   Ans. C

10. Modern evolutionary theory recognizes that evolution:

    A. moves toward a perfect state and towards greater complexity
    B. does not necessarily move towards a perfect state or greater complexity
    C. begins simple and progresses towards greater complexity and size
    D. tends to minimize complexity in higher life forms and maximize complexity in lower life forms

    Ans. B

11. The person that is credited with developing the idea that organisms pass on traits acquired during their lifetimes to their offspring is:

    A. Charles Darwin
    B. Leonardo da Vinci
    C. Aristotle
    D. Jean Baptiste de Lamarck

    Ans. D

12. Charles Darwin found that the actual mechanism of evolution is:

    A. mutation
    B. a change in the frequency of alleles
    C. natural selection
    D. a change in the frequency of DNA nucleotides

    Ans. C

13. Charles Darwin was appointed naturalist on the ship:

    A. H.M.S. Pentifore
    B. H.M.S. Beagle
    C. H.M.S. Galapagos
    D. H.M.S. Victoria

    Ans. B

14. When people selected certain desirable traits among domesticated plants and animals the process was known as:

    A. artificial selection
    B. selective breeding
    C. economic selection
    D. agricultural selection

    Ans. A

Darwin and Natural Selection

15. Charles Lyell was a scientist who studied:

    A. plants and plant ecosystems
    B. animals and animal natural history
    C. the earth and geological processes
    D. archeology and anthropology

    Ans. C

16. Thomas Malthus, an early contemporary of Darwin, was:

    A. a naturalist and science writer
    B. a geologist and archeologist
    C. a biologist and geologist
    D. a clergymen and economist

    Ans. D

17. Malthus wrote that population increases:

    A. arithmetically
    B. geometrically
    C. mathematically
    D. systematically or in precise patterns

    Ans. B

18. Malthus wrote that food resources increase:

    A. arithmetically
    B. geometrically
    C. systematically
    D. sporadically

    Ans. A

19. The relationship between the rate at which populations of consumers increase and the rate at which food increases is one in which:

    A. food supplies and populations increase proportionally
    B. populations increase faster than food supplies and outstrip food supplies
    C. food supplies increase at a faster rate than populations
    D. populations increase at a rate slightly less than food supplies due to their interdependence

    Ans. B

20. Malthus suggested that populations are regulated by the following:

    A. geological events and wars
    B. disease and war
    C. contraception and disease
    D. contraception and social strife

    Ans. B

21. As individuals within a population adapt to a new environment, and eventually accumulate adaptations that are favorably selected, this could result in:

    A. population equilibrium
    B. population decrease
    C. increasing selective pressure
    D. the origin of a new species

    Ans. D

22. Who is the individual that arrived at the conclusion that evolution occurred by natural selection, and presented his views with Charles Darwin at the Linnaean Society?

    A. Charles Lyell
    B. Thomas Malthus
    C. Alfred Wallace
    D. Jean Baptiste de Lamarck

    Ans. C

23. Overproduction, variation, limits on population growth, and survival to reproduce are all mechanisms of evolution by:

    A. induced selection
    B. artificial selection
    C. natural selection
    D. dynamic population evolutionary shifts

    Ans. C

24. Overpopulation does not occur among natural populations of organisms, because there are:

    A. limitations on how fast individuals within a population can reproduce
    B. limitation on their life-spans, which limit the rate at which they can reproduce
    C. feedback mechanisms that promote information to individuals within populations that cause them to arbitrarily not overproduce
    D. there are limiting factors that limit the size of populations

    Ans. D

Darwin and Natural Selection

25. Genetic variation within populations leads to the possibility that individuals improve their:

    A. survival and genetic equilibrium
    B. reproductive success and fitness
    C. reproductive success and survival
    D. fitness and gene variability

    Ans. C

26. Natural selection results in:

    A. increase in favorable alleles and decrease in unfavorable alleles
    B. initially an increase in favorable alleles that eventually attain equilibrium
    C. populations that steadily increase in size
    D. populations that eventually attain equilibrium

    Ans. A

27. Predation and disease are mechanisms that act as agents or forces of:

    A. evolution
    B. natural selection
    C. survival of the fittest
    D. adaptive fitness

    Ans. B

28. Geographical isolation of populations may result in:

    A. formation of a new species
    B. extinction of a species due to lack of population interaction
    C. extinction of a species due to lack of fitness
    D. a different mode of inheritance

    Ans. A

29. Darwin's theory of evolution and Mendelian genetics were united to form a unified explanation of evolution known as:

    A. Neo-Mendelian evolution
    B. genetic evolution
    C. synthetic theory of evolution
    D. Darwinian-Mendelian theory of evolution

    Ans. D

30. Mutation provides genetic variability that is acted upon by:

    A. adaptation
    B. fitness
    C. evolution
    D. natural selection

    Ans. D

31. The process of determining the order of nucleotide bases in a strand of DNA that codes for a gene shared by several organisms is referred to as:

    A. replication identification
    B. replication-transcription sequencing
    C. DNA sequencing
    D. nucleosome ordering

    Ans. C

32. Codons that code for specific amino acids are the same among all organisms, which indicates:

    A. that all organisms are identical in form
    B. that all organisms are identical in function
    C. that all life is related or has a common ancestry
    D. that all organisms have identical traits

    Ans. C

33. Wings of bats, flippers of dolphins, arms of humans, are examples of similar anatomical structures that have evolved from an evolutionary organ. Features similar in underlying structure in different species owing to their descent from a common ancestor are termed:

    A. analogous
    B. homologous
    C. synonymous
    D. evolutionary deviations

    Ans. B

Darwin and Natural Selection

34. Lungs in mammals, trachea insects, or the wings of birds and insects are similar in function but there is no evolutionary evidence that these organisms have evolved from a common ancestor. Organs like these that have similar functions in different organisms are referred to as:

    A. analogous
    B. homologous
    C. synonymous
    D. phylogenetic variants

    Ans. A

35. When an organ or body structure no longer has a selective advantage, or it loses some or all of its function, it is referred to as:

    A. an ancestral remnant
    B. a regressive organ or structure
    C. an evolutionarily extinct organ or structure
    D. a vestigial organ or structure

    Ans. D

36. When a new species evolves, it is adapted to certain physical conditions, and to certain kinds of biological interactions. The range where this new species has evolved is referred to as:

    A. evolutionary apex
    B. center of origin
    C. bio-geographical center of distribution
    D. maternal origin of habitat

    Ans. B

37. Explain the difference between evolution and natural selection, and how they interrelate.

38. Why is genetic variation so important to a species?

39. What is the difference between artificial selection and natural selection?

40. Explain why the concept of evolution does not agree with the present concept of evolution.

Chapter 18

Microevolution and Speciation

1. Concerning the evolutionary process, one can conclude that:

   A. it is a constant ongoing process
   B. it continued until humans evolved
   C. it continues throughout all of the lower forms of life, but is not really applicable to the higher forms of life
   D. it is the process that is responsible for the evolution of the five biological Kingdoms, but presently is only applicable to the microbes

   Ans. A

2. African prostitutes that appear to be resistant to the HIV-1 AIDS virus, biologically represent or demonstrate:

   A. they have evolved resistance to the HIV-1 AIDS virus
   B. they have evolved HLA proteins in response to their repeated exposure to HIV-1 AIDS virus
   C. they demonstrate that within a large population there is a large amount of genetic diversity and protein polymorphism
   D. they most probably have resistance because they select males that are not badly infected with HIV-1

   Ans. C

3. When one considers the mechanisms that lead to evolutionary damages within populations, such as AIDS-immune prostitutes, one of the primary requirements for evolution by natural selection is:

   A. a high rate of natural mutation due to exposure to mutagenic agents
   B. genetic variation within a population
   C. the population is capable of evolving characteristics or traits that act in response to life-threatening agents
   D. the recognition of drugs by researchers that evolve in response to mutation

   Ans. B

Microevolution and Speciation

4. A group of individuals that belong to the same species, and live in a particular place or area at a specific time, refers to what biological level of organization?

   A. class
   B. community
   C. family
   D. population

   Ans. D

5. Within a group of individuals that belong to the same species, and occupying the same region or area, there exists considerable genetic variation. The total genetic material of all of these individuals is referred to as:

   A. gene population
   B. gene pool
   C. gene summation
   D. genetic variant

   Ans. B

6. In diploid (2n) organisms, members of a chromosome pair are referred to as:

   A. paired chromosomes
   B. sister chromosomes
   C. homologous chromosomes
   D. parental chromosomes

   Ans. C

7. In diploid (2n) organisms, each individual within a given population possesses different combinations of:

   A. alleles
   B. chromosomes
   C. endosperm
   D. cells and tissues

   Ans. A

8. If the frequency of genes within a population remain constant from generation to generation, this indicates that this population:

   A. has attained dynamic equilibrium
   B. is on the verge of going extinct
   C. is well adjusted
   D. is not undergoing evolutionary change

   Ans. D

9. When small, gradual genetic changes occur within a population, it is referred to as:

   A. natural selection
   B. selective variation
   C. mutation
   D. microevolution

   Ans. D

10. Fossil evidence indicates the opossum is an animal that has not shown significant genetic change over the past several hundred thousand years. This evidence indicates that the opossum:

    A. possesses very little genetic variation
    B. is in genetic equilibrium
    C. has variation in its alleles
    D. is in adaptive stability

    Ans. B

11. The Hardy-Weinberg principle represents an ideal situation that describes the frequencies of alleles in the genotypes of an entire breeding population. The Hardy-Wienberg principle represents a situation that:

    A. frequently occurs in nature
    B. eventually represents the frequency of alleles for all populations
    C. probably never occurs in nature
    D. occurs in all population that are undergoing evolutionary changes

    Ans. C

12. The Hardy-Weinberg principle assumes that:

    A. mutation occurs at a constant rate
    B. mutation occurs at a very low frequency within populations
    C. no mutation occurs
    D. mutation is the primary force involved in maintaining genetic variation

    Ans. C

13. The Hardy-Weinberg principle assumes that:

    A. matings between genotypes occur in proportion to the frequencies of the genotypes
    B. mates select one another regarding preference or there is preferential mating
    C. offspring will frequently mate with one another
    D. offspring are never allowed to mate with one another

    Ans. A

Microevolution and Speciation

14. The Hardy-Weinberg principle assumes that:

    A. migration between populations is essential to maintain genetic balance
    B. no migration exists between populations
    C. migration between population is natural, therefore it is taken into account
    D. migration can not exceed a migration coefficient of 1.0

    Ans. B

15. The Hardy-Weinberg principle assumes that:

    A. no natural selection occurs
    B. natural selection is the foundation of the evolutionary process, and occurs at a specific rate for each species
    C. the effects of natural selection are too small or negligible to be accurately assessed
    D. natural selection is essential to the evolutionary process, and has the highest value

    Ans. A

16. The Hardy-Weinberg principle assumes that:

    A. population size is irrelevant
    B. population size must be small, in order to assess the changes
    C. population size must be intermediate or medium, because they represent populations that are stable and in equilibrium
    D. population size must be large so that the effect of changes is small

    Ans. D

17. The Hardy-Weinberg principle is able to predict:

    A. changes in population size
    B. changes in rates of mutation
    C. changes in the frequency of alleles
    D. changes in the frequency of males and females

    Ans. C

18. A random permanent change in DNA is referred to as:

    A. genetic modification
    B. DNA alteration
    C. mutation
    D. genetic variation

    Ans. C

19. Consider two populations, one with 20,000 individuals, and one with 20 individuals. If an allele occurs at a frequency of 5%, or 0.05, how many individuals within each population possess this allele?

A. 4,000 and 4 individuals
B. 5,000 and 5 individuals
C. 1,000 and 1 individual
D. 2,000 and 2 individuals

Ans. D

20. A particular allele is more likely to be lost:

A. within large populations
B. within small populations
C. when random mating occurs
D. when selective mating occurs

Ans. B

21. Production of random evolutionary changes in small breeding populations is known as:

A. genetic drift
B. differential genetic population shift
C. nonequilibrium genetic change
D. allele frequency displacement

Ans. A

22. The ultimate source of all new alleles is:

A. evolution
B. mutation
C. natural selection
D. non-random mating

Ans. B

23. Mutations that occur in the reproductive cells:

A. are usually temporary
B. are permanent
C. usually cause the organism that bears the mutation to become sterile
D. usually occur during the postproductive cycle

Ans. B

Microevolution and Speciation

24. The genetic drift that results from a small number of individuals from a large population colonizing a new area is referred to as:

A. founder effect
B. pioneer species
C. colonizer factor
D. differential allele variation

Ans. A

25. If a population experiences a considerable loss of population, that alters the frequency of alleles that existed among the population preceding the decline. This is referred to as:

A. differential gene flow
B. founder effect
C. genetic bottleneck
D. genetic survivor benefit

Ans. C

26. Mutations that occur in body cells or somatic cells

A. exhibit the greatest effect
B. are the most variable regarding their expression
C. are equal regarding their expression when compared to reproductive cells
D. are not heritable

Ans. D

27. Members of a species tend to be distributed in local populations that may be isolated genetically from other populations. The migration of individuals between populations results in the movement of alleles, that is referred to as:

A. genetic shift
B. gene migration
C. gene flow
D. DNA differential migration

Ans. C

28. When migration of individuals occurs between two populations that have been isolated genetically, one would expect these populations to:

    A. retain most of the traits found in the preceding population
    B. become more similar genetically
    C. resist the forces of migration because they are adapted to specific selective agents
    D. remain isolated, because they have evolved characteristics that are not compatible with other populations

    Ans. D

29. What factor tends to check the random effects of genetic drift, mutation, and gene flow?

    A. natural selection
    B. evolution
    C. adaptation
    D. founder effect

    Ans. A

30. Over successive generations, the proportion of more favorable alleles increases within a population. This process is referred to as:

    A. evolutionary process
    B. gene flow
    C. genetic drift
    D. natural selection

    Ans. D

31. Three kinds of selection occur that cause changes in the normal distribution of phenotypes in a population. These three types of selection are:

    A. natural selection, artificial selecting, and differential selecting
    B. natural selection, artificial selection, and environmental selection
    C. directional selection, disruptive selection, and stabilizing selection
    D. disruptive selection, directional selection, and natural selection

    Ans. C

32. A group of more or less distinct organisms capable of interbreeding with one another in nature but reproductively isolated from one another, refers to:

    A. a biological family
    B. a population
    C. a community
    D. a species

    Ans. D

Microevolution and Speciation

33. Reproductive isolating mechanism, temporal isolation, behavioral isolation, and mechanical isolation are all processes that refer to mechanisms that prevent interbreeding between different:

    A. genetic isolates
    B. population
    C. species
    D. breeds

    Ans. C

34. The presence in a population of two or more alleles for a given gene, is referred to as:

    A. genetic variation
    B. genetic polymorphism
    C. genetic drift
    D. genetic selection

    Ans. B

35. Heterozygous advantage is demonstrated by the selective advantages bestowed on heterozygous carriers. In humans such an advantage is bestowed upon individuals that carry the allele for:

    A. sickle-cell anemia
    B. malaria
    C. Parkinson's disease
    D. Turner's syndrome

    Ans. A

36. The mechanism of natural selection results in:

    A. the development of a "perfect" organism
    B. offspring that are superior to their parents
    C. eliminating or "weeding out" those phenotypes that are less adaptive
    D. decreasing the frequency of all alleles, due to the random nature of natural selection

    Ans. C

37. Differentiate between the following:
    a. natural selection
    b. stabilizing selection
    c. directional selection
    d. disruptive selection

38. Explain how the following mechanisms function in preventing interbreeding between species:
    a. reproductive isolation
    b. temporal isolation
    c. behavioral isolation
    d. mechanical isolation

39. Differentiate between allopatric speciation and symatric speciation.

40. Differentiate between the following:
    a. gene flow
    b. mutation
    c. genetic drift
    d. genetic bottlenecks

Chapter 19

Macroevolution and the History of Life

1. One of the most serious threats to the survival of many species today is:

   A. the changing weather patterns
   B. habitat disruption or loss of habitats
   C. the problem associated with the ever-increasing use of nuclear energy
   D. pollution that is generated by humans

   Ans. B

2. Extinctions have always occurred, but the current extinction episode is occurring:

   A. with greater severity to the organisms involved
   B. at a slower rate due to the organisms involved
   C. at the same rate at which extinction has always occurred
   D. at a faster rate

   Ans. D

3. Presently, biologists estimate that at least one species becomes extinct every:

   A. minute
   B. hour
   C. day
   D. week

   Ans. C

4. At the present rate of extinction, it is estimated that a substantial portion of Earth's biological diversity will be eliminated within the:

   A. next five years
   B. next few decades
   C. next few centuries
   D. next thousand years

   Ans. B

5. Clear-cutting of forest land, overgrazing of domestic livestock, draining of marshes and wetlands, and construction of dams, are all classified as agents that:

   A. cause a loss of habitat
   B. are not economically sound
   C. provide new habitat to a greater number of species
   D. enhance an area for recreation

   Ans. A

6. The dusky seaside sparrow became extinct in 1987, due to:

   A. overhunting, especially by young teenagers
   B. poisoning of their food by pollution
   C. disease that was introduced from exotic birds brought into this country from Central America
   D. a loss of habitat

   Ans. D

7. Major evolutionary events that occur in groups of species over geological time are referred to as:

   A. natural selection
   B. evolution
   C. macroevolution
   D. adaptive radiation

   Ans. C

8. If an organism succeeded because it possessed a characteristic or trait that some other individuals within the population did not possess, it would have:

   A. a greater success ratio
   B. a selective advantage
   C. an increased adaptive coefficient
   D. evolved at a faster rate

   Ans. B

9. The evolution of many related species from one or two ancestral species in a relatively short period of time, is referred to as:

   A. adaptive radiation
   B. ecological adaptiveness
   C. ecological niche
   D. macroevolution

   Ans. A

10. New ecological roles or ways of living that were previously not utilized by an ancestral organism, are referred to as:

    A. adaptive radiation
    B. ecological fitness
    C. ecological adaptiveness
    D. adaptive zone

    Ans. D

11. An organism's role or its "occupation" within a biological community, can only be occupied by one species due to its adaptive zone. This unique adaptive role is referred to as:

    A. ecological fitness
    B. ecological adaptiveness
    C. ecological niche
    D. adaptive radiation

    Ans. C

12. Adaptive radiation appears to be more common during periods of:

    A. environmental equilibrium
    B. major environmental change
    C. prior to an extinction of a species
    D. continuous mild weather

    Ans. B

13. There are two types of extinction that are recognized. They are:

    A. low frequency extinction and high frequency
    B. minor levels of extinction and major levels of extinction
    C. background extinction and mass extinction
    D. nonadaptive extinction and non-random extinction

    Ans. C

14. There are two different models that have been developed to explain evolutionary change. One model that has been proposed questions whether the fossil record is as incomplete as it initially appeared; it proposes that the fossil record accurately reflects evolution as it occurs. This model is referred to as:

    A. punctuated equilibrium
    B. gradualism
    C. microevolution
    D. constant equilibrium

    Ans. A

Macroevolution and the History of Life

15. Periods of time when there is no evolutionary change are referred to as:

    A. stasis
    B. static periods
    C. periods of equilibrium
    D. periods of homogeneity

    Ans. A

16. There are two different models that have been developed to explain evolutionary change. This model proposes that evolution proceeds at a more or less steady rate but is not observed in the fossil record because the record is incomplete. This model is referred to as:

    A. punctuated equilibrium
    B. gradualism
    C. microevolution
    D. constant equilibrium

    Ans. B

17. Periods of time when there is no evolutionary change could result from:

    A. directional selection
    B. natural selection
    C. niche separation followed by niche equilibrium
    D. stabilizing selection

    Ans. D

18. Evolutionary biologists and evolutionary ecologists generally agree that the primary mechanism responsible for evolution is:

    A. adaptive radiation
    B. ecological niche
    C. natural selection
    D. microevolution

    Ans. C

19. Most scientists generally accept the hypothesis that life developed from:

    A. simpler life forms that no longer exist
    B. nonliving matter
    C. life transported to earth from an extinct portion of the universe
    D. life that previously existed prior to the formation of our present universe

    Ans. B

20. The atmosphere of the early Earth or pre-biotic Earth, contained little or no:

    A. carbon dioxide (CO_2)
    B. water (H_2O)
    C. free hydrogen (H_2)
    D. free oxygen (O_2)

    Ans. D

21. Initially, small organic molecules formed spontaneously and accumulated over time. Eventually these small organic molecules bonded together to form macromolecules which eventually formed molecular assemblages that developed into cell-like structures. These macromolecular assemblages succeeded or failed as a result of:

    A. natural selection
    B. directional selection
    C. stabilizing selection
    D. disruptive selection

    Ans. A

22. The Earth's early atmosphere or the prebiotic Earth's atmosphere was strongly:

    A. oxidizing
    B. reducing
    C. ion free
    D. radiation free

    Ans. B

23. The four requirements for the origin of life are:

    A. no free oxygen, energy, available chemicals or compounds, and time
    B. water, energy, available chemicals or compounds, and heat
    C. water, carbon dioxide, energy, and cosmic radiation
    D. water, energy, ionizing radiation, and cold temperatures to cause condensation of molecules

    Ans. A

24. In Miller and Urey's experiment that they designed to simulate conditions similar to those prevalent on the early Earth they produced:

    A. living prokaryotic-like cells
    B. DNA and RNA
    C. amino acids and other organic compounds
    D. proteins and other molecular organic compounds

    Ans. C

25. The Earth is approximately:

    A. 2,000 years old
    B. 4.6 million years old
    C. 4.6 billion years old
    D. 4.6 trillion years old

    Ans. C

26. Based on scientific evidence, most scientists think it is more likely that organic polymers formed and accumulated:

    A. in primordial seas as a "sea of organic soup"
    B. on the bottoms of very deep seas where the pressure was very intense, thus causing organic compounds to bond together
    C. in rain drops that accumulated ions and micromolecules from the atmosphere
    D. on rock and clay surfaces

    Ans. D

27. Scientists have been able to synthesize several prebiotic assemblages of organic polymers that resemble simple life forms. These assemblages are referred to as:

    A. prekaryotes
    B. protobionts
    C. prokaryotes
    D. stromatolites

    Ans. B

28. Fossil evidence indicates that cells existed and were thriving:

    A. 2,000 years ago
    B. 3.5 million years ago
    C. 35 million years ago
    D. 3.5 billion years ago

    Ans. D

29. Rocklike columns composed of many minute layers of prokaryotic cells form reefs that are referred to as:

    A. protobionts
    B. reefoids
    C. stromatolites
    D. protofossilites

    Ans. C

30. The earliest cells that evolved required nutrients to survive. These early cells were probably:

    A. heterotrophs
    B. autotrophs
    C. aerobic
    D. facultative parasites

    Ans. A

31. The first photosynthesizers probably used the energy of sunlight to split hydrogen-rich molecules like:

    A. water (H_2O)
    B. hydrogen sulfide (H_2S)
    C. methane (CH_4)
    D. ammonia (NH_3)

    Ans. B

32. The first photosynthetic autotrophs to split water in order to release hydrogen were the:

    A. cyanobacteria
    B. prokaryotes
    C. green algae
    D. diatoms

    Ans. A

33. Eukaryotes evolved from what group of organisms?

    A. endosymbionts
    B. cyanobacteria
    C. prokaryotes
    D. sulfur bacteria

    Ans. C

34. Organisms that evolved respiratory pathways that utilize free oxygen (O_2) are referred to as:

    A. autotrophic oxygenators
    B. endosymbionts
    C. anaerobes
    D. aerobes

    Ans. D

35. The theory that suggests that eukaryotic organelles such as mitochondria and chloroplasts may have originated from a mutualistic relationship between two prokaryotes is referred to as:

A. endosymbiont theory
B. eukaryotic symbosis
C. eukaryotic germ theory
D. eukaryotic microevolution

Ans. A

36. Which of the four geological methods for classifying time is the largest unit?

A. eon
B. epoch
C. era
D. period

Ans. C

37. Explain how eukaryotic cells most probably evolved.

38. Explain how the first cells most likely developed from nonliving matter.

39. Differentiate between punctuated equilibrium and graduation.

40. Explain why adaptive radiation appears to be more common during periods of major environmental change.

Chapter 20

The Classifications of Organisms

1. The greatest amount of species diversity exists in:

   A. caves
   B. seas
   C. swamps
   D. tropical rain forests

   Ans. D

2. About how many species of organisms have been identified?

   A. 1 million
   B. 2 million
   C. 5 million
   D. 10 million

   Ans. B

3. The greatest contributing factor that has lead to a decrease in species diversity has been:

   A. loss of habitat
   B. disease and parasitism
   C. human-induced pollution
   D. increased use of nuclear energy

   Ans. A

4. Two programs that are presently attempting to assess biological diversity are:

   A. biological and Bioassess
   B. RAG and BIOTECH
   C. BIODAD and BIOSAV
   D. RAP and BIOTROP

   Ans. D

The Classifications of Organisms

5. Four groups of organisms are being extensively studied in this assessment program. The four that are being studied are:

   A. plants, animals, fungi, and bacteria
   B. plants, animals, algae, and bacteria
   C. woody plants, vertebrates, ants, and butterflies
   D. woody and herbaceous plants, birds and mammals, insects and spiders, and fungi

   Ans. C

6. In rain forests it has long been known that many mammals are dependent upon plants as a source of food and shelter. Rain forest plants are dependent upon mammals to:

   A. pollinate their flowers
   B. disperse their seeds
   C. aid in the maturation of their fruit
   D. remove flowers that were not pollinated

   Ans. B

7. Who was the botanist and natural historian that simplified scientific classification of organisms?

   A. Escherich
   B. Jenner
   C. Linnaeus
   D. Levey

   Ans. C

8. Before the scientific classification of organisms was simplified, each organism had a lengthy descriptive name, sometimes composed of ten or more:

   A. German words
   B. Greek words
   C. French words
   D. Latin words

   Ans. D

9. The present system, that was developed by an 18th century botanist and natural historian, is referred to as:

   A. binomial system of nomenclature
   B. binomial system of taxonomy
   C. binomial system of phyletic nomenclature
   D. the universal system of organismic classification

   Ans. A

10. Each species is assigned a name that consists of:

A. two words
B. three words
C. four words
D. up to six words

Ans. A

11. The specific epithet refers to the

A. genus name
B. species name
C. taxon name
D. phyletic name

Ans. B

12. The genus and species names for an organism always follow these guidelines:

A. genus and species names are both capitalized, and both are underlined
B. genus and species are both capitalized, and the genus name is underlined
C. genus name is capitalized and the species name is not, and both are underlined
D. genus name is capitalized and the species name is not capitalized, while the genus name is double underlined and the species name is single underlined

Ans. C

13. The genus and species names are:

A. specific for each organism within a particular Kingdom, but a plant that belongs to the plant kingdom could have the same genus and species name as an animal that belongs to the animal kingdom
B. are specific for each eukaryotic organism, but bacteria that belong to the kingdom Monera occasionally have the same genus and species names
C. are specific for each organism, but the system is flawed, because it varies from one country to another
D. are specific for each organism, and are universally accepted throughout the entire international scientific community

Ans. D

14. A taxonomic grouping at any level, is referred to as:

A. a classifier
B. a taxon
C. a specific biological grouping
D. a systemic group

Ans. B

The Classifications of Organisms

15. A group of closely related species are assigned to the same:

    A. class
    B. order
    C. family
    D. genus

    Ans. D

16. A group of phyla are grouped into a taxonomic group called a:

    A. class
    B. order
    C. Kingdom
    D. family

    Ans. C

17. When classifying plants and fungi, the term phylum is replaced by:

    A. Plantae and Fungi
    B. phyla
    C. division
    D. taxon

    Ans. C

18. Some biologists use a level of classification above the Kingdom, called a domain, which divides:

    A. prokaryotes and eukaryotes
    B. prokaryotes, viruses, and eukaryotes
    C. autotrophs and heterotrophs
    D. archaebacteria and eubacteria

    Ans. A

19. The only one of the taxa that actually exists in nature is the:

    A. genus
    B. species
    C. organism
    D. population

    Ans. B

20. When taxonomists ignore minor variations and group organisms into already existing taxa, the practice is referred to as:

A. taxon synthesis
B. taxonomic condensation
C. inclusive taxonomic classification
D. lumping

Ans. D

21. When taxonomists subdivide taxa on the basis of major differences, establishing separate categories for forms that do not fall naturally into one of the existing classifications, the practice is referred to as:

A. taxonomic division
B. inclusive systemic division
C. splitting
D. taxonomic subdivision and separation

Ans. C

22. Populations inhabiting different geographic areas often display certain consistent characteristics or traits that distinguish them from other populations of the same species. These unique populations are referred to as:

A. a gradient population
B. a breed
C. an inclusive taxonomic population
D. a subspecies

Ans. D

23. Most biologists recognize how many taxonomic Kingdoms?

A. two
B. four
C. five
D. six

Ans. C

24. The classification of organisms into groups is to evolutionary relationships, as:

A. systematics is to phylogeny
B. evolution is to taxonomy
C. plant is to animal
D. common ancestry is to organism

Ans. A

The Classifications of Organisms

25. Taxa that share the same common ancestor is to taxa that do not include a common ancestor, as:

    A. polyphyletic is to monophyletic
    B. multiple ancestry is to single ancestry
    C. monophyletic is to polyphyletic
    D. taxon is to clade

    Ans. C

26. A common ancestor is to all the taxa descended from it, as:

    A. systematics is to taxon
    B. polyphyletic is to clade
    C. polyphyletic is to monophyletic
    D. taxon is to clade

    Ans. D

27. Homologous structures are to analogous structures as:

    A. bat wings and butterfly wings are to beetle wings and bird wings
    B. bat wings and bird wings are to butterfly wings and beetle wings
    C. fish fins are to butterfly legs as shark fins are to ant legs
    D. lizard scales are to insect scales as bird feathers are to clam shells

    Ans. B

28. Phenetics, cladistics, and classical evolutionary taxonomy are the three major approaches to:

    A. evolutionary biology
    B. homology
    C. classification
    D. understanding molecular clocks

    Ans. C

29. Traits that have essentially remained unchanged throughout a group of species, are referred to as:

    A. ancestral characters
    B. taxonomic characters
    C. analogous taxonomic traits
    D. branching characteristics

    Ans. A

30. A trait or group of traits not present in ancestral species because they evolved more recently, are referred to as

A. chronological traits
B. divergent evolution
C. divergent traits
D. derived characters

Ans. D

31. Beaks, feathers, and no teeth are to bird as:

A. gills, scales and fins are to porpoises
B. breathe air, nurse young, and maintain constant body temperatures are to porpoises
C. breathe air, nurse young, and cold blooded body temperatures are to porpoises
D. breathe air, scales, and fins are to porpoises

Ans. B

32. Taxonomists that use numerical taxonomy are referred to as:

A. evolutionary taxonomists
B. cladists
C. pheneticists
D. analytical taxonomists

Ans. C

33. When differences in nucleotide sequence of DNA or amino acid sequence of proteins in two taxonomic groups branched off from a common ancestor, the specific genes and specific proteins can be used as:

A. nucleotide-protein analysis
B. specific tracers
C. molecular differentiators
D. molecular clocks

Ans. D

34. Shared derived characters indicate a more recent common ancestor than:

A. shared ancestral characters
B. characteristics of siblings
C. characteristics of parents and siblings
D. characteristics of a species

Ans. A

The Classifications of Organisms

35. The correct hierarchical system of classification used includes:

    A. Kingdom, phylum, order, family, class, genus, and species
    B. Kingdom, phylum, class, order, family, genus, and species
    C. Kingdom, phylum, family, order, class, genus, and species
    D. Kingdom, order, phylum, class, family, genus, and species

    Ans. B

36. How many of the Kingdoms of classification possess eukaryotic organisms?

    A. 1
    B. 2
    C. 4
    D. 5

    Ans. C

37. Describe the binomial system of nomenclature.

38. What are the advantages of having five Kingdoms, as opposed to the original concept of two Kingdoms?

39. Compare phenetic, cladism, and classical evolutionary approaches to taxonomy.

40. How do taxonomists use nucleotide and amino acid sequences in determining taxonomic relationships?

Chapter 21

Microorganisms: Viruses, Bacteria, and Protists

1. One reason viruses are considered nonliving entities is because they :

   A. can reproduce only after they have infected a host cell
   B. can make several copies of themselves at once
   C. infect only animals
   D. are shaped differently from other forms of life

   Ans. A

2. The outer structure (coat) of a virus is mostly composed of:

   A. carbohydrates
   B. lipids
   C. proteins
   D. nucleic acids

   Ans. C

3. The coat of a virus is known as a:

   A. capsule
   B. capsomere
   C. centromere
   D. capsid

   Ans. D

4. A virus that can infect only bacteria is known as a:

   A. bacterioviroid
   B. bacteriophage
   C. bacterioprion
   D. bacteriovirino

   Ans. B

5. Place the following steps of viral reproduction in order from beginning to end:
   1. Assembly
   2. Attachment
   3. Penetration
   4. Release
   5. Replication

   A. 3-1-2-4-5
   B. 5-2-4-3-1
   C. 2-3-5-1-4
   D. 4-5-3-1-2

   Ans. C

6. One characteristic that sets RETROVIRUSES apart from other viruses is that they must make __________ before they can replicate themselves inside a host cell.

   A. DNA from proteins
   B. DNA from RNA
   C. RNA from proteins
   D. RNA from DNA

   Ans. B

7. Bacteria are __________ forms of life.

   A. prokaryotic
   B. eukaryotic
   C. protistan
   D. fungal

   Ans. A

8. Prokaryotes differ from eukaryotes in that they do not contain __________ as a component of their cells.

   A. DNA
   B. RNA
   C. organelles
   D. proteins

   Ans. C

9. Where would you expect to find PEPTIDOGLYCAN as it would be associated with bacteria?

   A. around the outside of the cell
   B. in the cytoplasm
   C. in their mitochondria
   D. in the nucleus

   Ans. A

10. A polysaccharide-rich mucoid coating that is often found surrounding the cells of pathogenic bacterial species is called a/an

A. capsomere
B. capsule
C. capsid
D. centromere

Ans. B

11. A highly-resistant survival structure that can be formed by some bacteria (in response to adverse growth conditions in their environment) is called a/an:

A. exospore
B. haplospore
C. diplospore
D. endospore

Ans. D

12. Bacteria reproduce by:

A. binary fission
B. budding
C. fragmentation
D. sexual reproduction

Ans. A

13. Most bacteria can be described as _________ with regard to their food and energy supplies as they cannot make their own food.

A. autotrophs
B. heterotrophs
C. lithotrophs
D. phototrophs

Ans. B

14. Bacteria fill the niche of _________ in most ecosystems.

A. producer
B. photosynthesizer
C. decomposer
D. primary consumer

Ans. C

15. Some bacteria form symbiotic relationships with legume plants because the bacteria have the ability to fix what element from the gases in the atmosphere?

    A. oxygen
    B. hydrogen
    C. helium
    D. nitrogen

    Ans. D

16. The two groupings within types of bacteria are:

    A. eubacteria and archaeobacteria
    B. archaeobacteria and eukaryota
    C. eubacteria and prokaryota
    D. prokaryota and eukaryota

    Ans. A

17. A thermophilic bacterium loves:

    A. low salt concentrations
    B. high pH conditions
    C. high temperatures
    D. low pressures

    Ans. C

18. One of the most conspicuous differences between archaeobacteria and other bacteria is:

    A. the total lack of a cell wall
    B. the lack of peptidoglycan from their cell wall
    C. the presence of a double plasma membrane
    D. the complete absence of a plasma membrane

    Ans. B

19. The scientific term BACILLUS describes a bacterium that is:

    A. spiral
    B. helical
    C. spherical
    D. rod shaped

    Ans. D

20. The scientific term COCCUS describes a bacterium that is:

    A. spiral
    B. helical
    C. spherical
    D. rod shaped

    Ans. D

21. Which group of bacteria lack a cell wall of any kind?

    A. mycoplasmas
    B. gram-positives
    C. gram-negatives
    D. archaeobacteria

    Ans. A

22. In which bacterial group would you expect to find a thick peptidoglycan cell wall?

    A. archaeobacteria
    B. mycoplasmas
    C. gram-negatives
    D. gram-positives

    Ans. D

23. *Escherichia coli*, the causative agent of several outbreaks of bloody diarrhea (obtained from eating undercooked contaminated meats), is classified in the group of ________ since it normally lives in the intestines.

    A. cyanobacteria
    B. enterobacteria
    C. nitrogen-fixing bacteria
    D. methanogenic bacteria

    Ans. B

24. Yogurt is formed by adding what type of bacteria to milk?

    A. cyanobacteria
    B. enterobacteria
    C. thermophilic bacteria
    D. lactic acid bacteria

    Ans. D

25. Which of these terms is not used to describe the shape of a typical bacterial cell?

    A. angular
    B. bacillus
    C. coccus
    D. spirilla

    Ans. A

26. Members of the kingdom ________ are among the simplest of eukaryotes.

    A. Fungi
    B. Protista
    C. Animalia
    D. Plantae

    Ans. B

27. Most protists live in what type of environment?

    A. dry
    B. hot
    C. high-altitude
    D. aquatic

    Ans. D

28. An example of a protist that lives on land, "crawls" over rotting logs and leaf litter, and can produce spores by meiosis is a/an:

    A. slime mold
    B. paramecium
    C. dinoflagellate
    D. euglena

    Ans. A

29. Algae were once classified in the kingdom ________, because of structural similarities, before being classified as a Protistan.

    A. Prokaryota
    B. Fungi
    C. Plantae
    D. Animalia

    Ans. C

30. A staple food plant that was infected by water molds, which destroyed the crop and caused the starvation of many Irish people during the mid-1800's, was the _________ plant.

A. tomato
B. potato
C. carrot
D. lettuce

Ans. B

31. Into which nutritional classification would you classify algae?

A. chemolithotroph
B. protobiotroph
C. organoheterotroph
D. photoautotroph

Ans. D

32. The organism that is associated with "red tides," a phenomenon resulting from a bloom of algal organisms that leads to massive fish kills and can poison humans, is a/an:

A. paramecium
B. amoeba
C. dinoflagellate
D. water mold

Ans. C

33. Kelps (and some giant seaweeds) are a type of _________ algae.

A. yellow
B. green
C. brown
D. red

Ans. C

34. A word that is used to describe some types of protistans as the "first animals" is:

A. protozoa
B. phytoplankton
C. diatomic
D. prokaryotae

Ans. A

35. A "foraminiferan" is a _________ protistan.

    A. spongy
    B. ciliated
    C. shell-forming
    D. flagellated

    Ans. C

36. A common protozoan that engulfs its nutrients by surrounding the materials with a structure known as a "pseudopod" (false foot) is a/an:

    A. foraminiferan
    B. paramecium
    C. sporozoan
    D. amoeba

    Ans. D

37. A common protozoan that swims in most ponds and aquariums using small hair-like projections called "cilia" is a/an:

    A. foraminiferan
    B. paramecium
    C. sporozoan
    D. amoeba

    Ans. D

38. One way of cleaning up toxic chemical spills is by a process of "bioremediation." How does bioremediation work? What are some advantages (or disadvantages) to using bioremediation as opposed to usinq more conventional clean-up methods?

39. Describe the differences and similarities that would be seen between bacteria and viruses. Include in your discussion characteristics such as cell structure, sizes, shapes, modes of nutrition, reproductive strategies, etc.

40. Let's say you have chosen to debate the thought that "Bacteria do nothing for us but cause disease." You know this is an untrue statement, and it is up to you to prove it false. What facts could you supply to someone with the above belief to try to convince them that all bacteria are not bad? Remember, the more supporting facts you have, the stronger your argument!

41. It is believed that multicellular structure evolved within the kingdom Protista, and then led to the evolution of other multicellular organisms. Discuss how such a complex process can be related to such simple organisms.

Chapter 22

Fungal Life

1. Molds and yeast belong to the kingdom:

   A. Monera
   B. Protista
   C. Fungi
   D. Plantae

   Ans. C

2. Fungi fill the niche of ________ in ecosystems

   A. autotrophs
   B. producers
   C. consumers
   D. decomposers

   Ans. D

3. Fungi are best described as:

   A. eukaryotic
   B. prokaryotlc
   C. abiotic
   D. autotrophic

   Ans. A

4. The primary difference between a mold and a yeast is:

   A. nutritional strategies
   B. cell structure
   C. sensitivity to oxygen
   D. diseases caused by each

   Ans. B

Fungal Life

5. The most common reproductive structure formed by the multicellular molds is a:

   A. bud
   B. hyphae
   C. limb
   D. spore

   Ans. D

6. The thread-like structures that often make up the body of a mold is known as a:

   A. fruiting body
   B. stalk
   C. hyphae
   D. bud

   Ans. C

7. The "sac fungi" are classified in the taxonomic division of:

   A. ascomycota
   B. basidiomycota
   C. zygomycota
   D. deuteromycota

   Ans. A

8. The "imperfect fungi" are classified in the taxonomic division of:

   A. ascomycota
   B. basidiomycota
   C. zygomycota
   D. deuteromycota

   Ans. D

9. The "club fungi" are classified in the taxonomic division of:

   A. ascomycota
   B. basidiomycota
   C. zygomycota
   D. deuteromycota

   Ans. B

10. The bread molds are classified in the taxonomic division of:

    A. ascomycota
    B. basidiomycota
    C. zygomycota
    D. deuteromycota

    Ans. C

11. Truffles and edible morel mushrooms are classified in the same division with:

    A. most yeasts
    B. bread molds
    C. imperfect fungi
    D. wheat rusts

    Ans. A

12. Fungi reproduce

    A. only asexually
    B. only sexually
    C. with both sexual and asexual processes
    D. only through mitosis

    Ans. C

13. A fungus that has a cap, stalk, and gills is commonly known as:

    A. mushroom
    B. puff ball
    C. wheat rust
    D. bracket fungus

    Ans. A

14. Hyphae that have only one nucleus per cell are said to be:

    A. dikaryotic
    B. dizygous
    C. monozygous
    D. monokaryotic

    Ans. D

Fungal Life

15. When hyphae fuse, and two distinct nuclei can be seen in the cells, this condition is described as:

    A. monozygous
    B. dizygous
    C. dikaryotic
    D. monokaryotic

    Ans. B

16. The greatest biomass of a mushroom seen growing above the ground would be found:

    A. as its spores
    B. as its stalk
    C. under the surface of the soil
    D. as the fruiting body

    Ans. C

17. *Rhizopus nigrans* is the scientific name for:

    A. black bread mold
    B. wine-making yeast
    C. bread-making yeast
    D. morel mushrooms

    Ans. A

18. Compact "buttons" of hyphae are structures most closely associated with:

    A. yeasts
    B. bread mold
    C. fruit mildew
    D. mushrooms

    Ans. D

19. The fungi that are classified as members of deuteromycota are placed in this division because they:

    A. look like typical mushrooms
    B. are photosynthetic
    C. have no distinct sexual stages
    D. are the only fungi that cause plant disease

    Ans. C

20. A lichen is an association between

A. a fungus and a bacterium
B. a fungus and an algae
C. a fungus and a plant
D. two different fungi living together

Ans. B

21. The type of relationship exhibited by the organisms that compose a lichen is usually described as:

A. parasitism
B. mutualism
C. saprophytism
D. commensualism

Ans. B

22. A slow-growing "patch" of lichen that is about a meter in diameter may be as much as

A. 1000 years old
B. 100 years old
C. 10 years old
D. 1 year old

Ans. A

23. Lichens usually reproduce by

A. budding
B. binary fusion
C. sporulation
D. fragmentation

Ans. D

24. Most fungi are described as __________, which means they live on dead materials.

A. parasites
B. commensuals
C. saprophytes
D. pathogens

Ans. C

Fungal Life

25. Yeasts are classified in the taxonomic division of:

    A. Zygomycota
    B. Deuteromycota
    C. Basidiomycota
    D. Ascomycota

    Ans. D

26. A fungus that is used to make bread rise would be:

    A. *Saccharomyces cerevesiae*
    B. *Aspergillis fumigatis*
    C. *Candida albicans*
    D. *Penicillium notatum*

    Ans. A

27. A fungus that can cause mouth or vaginal infections is:

    A. *Saccharomyces cerevesiae*
    B. *Penicillium notatum*
    C. *Candida albicans*
    D. *Aspergillis fumigatis*

    Ans. C

28. A fungus that can produce a powerful antibiotic is:

    A. *Aspergillis tamarii*
    B. *Penicillium notatum*
    C. *Amanita virosa*
    D. *Rhizopus nigrans*

    Ans. B

29. Ringworm is caused by a:

    A. bacterium
    B. protozoan
    C. fungus
    D. worm

    Ans. C

30. Fungi were once considered to be plants because of their resemblance to plants, but their lack of _________ sets them apart from plants.

A. mitochondria
B. nuclei
C. chlorophyll
D. cell membrane

Ans. C

31. Fungi are found in what environments?

A. nearly all environments
B. only dark environments
C. only lighted environments
D. only warm environments

Ans. A

32. One environmental element that all fungi must have to live is:

A. warm temperature
B. water
C. light
D. soil

Ans. B

33. Complete the following analogy----Fungi : Chitin as

A. Protozoans : Cytoplasm
B. Bacteria : Nucleus
C. Animal cells : Peptidoglycan
D. Plants : Cellulose

Ans. D

34. Mycorrhizae are _________ fungal structures.

A. root-like
B. stalk-like
C. leaf-like
D. spore-like

Ans. A

Fungal Life

35. Approximately _________ of plant species have a symbiotic relationship with mycorrhizal fungi.

    A. 30%
    B. 50%
    C. 70%
    D. 90%

    Ans. D

36. Mycorrhizal fungi supply growth factors to symbiotic plants by:

    A. producing acids necessary to raise pH
    B. producing oxygen
    C. decomposing available organic matter
    D. making glucose by photosynthesis

    Ans. C

37. Give several examples of how fungi are of great economic importance. Include in your discussion examples of some processes and/or products that fungi are used for.

38. In the last several years fungi were discovered (one in Wisconsin, and another in Washington state) that have been described as, quite possibly, the "largest living organisms" on Earth, and yet only a few "toad stools" could be seen growing out of the ground. What do you suppose was meant by the above statement if so little of the fungus could be seen at that time?

39. Fungi are described as being extremely important to most ecosystems. Describe factors that make them so important to other living organisms found as cohabitants of any given ecosystem (i.e., what would be some possible effects of not having fungi in certain ecosystems).

40. Fungal species are disappearing from the world at an alarming rate, even though legislation like the "Endangered Species Act" is supposed to be protecting all organisms threatened with extinction. Discuss some reasons why so many fungal species are being lost, and some possible ways to prevent such an ecological tragedy.

Chapter 23

Plant Life

1. One of the unifying characteristics of almost all plants is:

   A. they possess roots and leaves
   B. their mode of nutrition is photosynthesis
   C. they possess green leaves and have stems
   D. they produce flowers which eventually form seeds

   Ans. B

2. Nutritionally, plants are capable of:

   A. absorbing chemicals and converting them to energy
   B. absorbing chemicals and converting them into organic compounds
   C. converting inorganic chemical energy into organic energy
   D. absorbing radiant energy, and converting it into chemical energy

   Ans. D

3. Plants are classified as:

   A. simple autotrophs
   B. facultative autotrophs
   C. photosynthetic autotrophs
   D. obligative autotrophs

   Ans. C

4. Photosynthetic pigments are to accessory photosynthetic pigments as:

   A. chlorophylls a and b are to carotenoids
   B. carotenoids are to chlorophylls a and b
   C. chlorophylls a and b are to xanthophylls
   D. chloropylls c and d are to cyanophylls

   Ans. A

5. Common ancestor is to descendent as:

   A. plant is to green algae
   B. green algae is to photosynthetic bacteria
   C. moss is to vascular plants
   D. green algae is to plants

   Ans. D

6. Important terrestrial adaptation (land adaptation) is to preventing drying out as:

   A. stroma is to reduction in evaporation
   B. cuticle is to reduction in evaporation
   C. chloroplast is to reduction in evaporation
   D. xylem is to reduction in evaporation

   Ans. B

7. Haploid generation is to diploid generation as:

   A. sporophyte generation is to gametophyte generation
   B. monocot generation is to sporophyte generation
   C. gametophyte generation is to sporophyte generation
   D. polyploid generation is to monoploid generation

   Ans. C

8. Fertilization by water is to fertilization by wind as:

   A. moss is to conifer
   B. moss is to fern
   C. moss is to flowering plant
   D. conifer is to flowering plant

   Ans. A

9. Nonvascular plants are to vascular plants as:

   A. gymnosperms are to flowering plants
   B. ferns are to flowering plants
   C. gymnosperms are to bryophytes
   D. bryophytes are to flowering plants

   Ans. D

10. What group of organisms can photosynthesize in addition to algae and plants?

    A. certain fungi
    B. certain bacteria
    C. certain corals
    D. certain sponges

    Ans. B

11. Gas exchange in stems and leaves, which permits photosynthesis to occur is accomplished by:

    A. the cuticle
    B. the lenticels
    C. the stomata
    D. the xylem and phloem combined

    Ans. C

12. When plants have a distinct haploid stage and diploid stage, this is referred to as:

    A. asexual and sexual generations
    B. binary generations
    C. mitotic and meiotic generations
    D. alternation of generations

    Ans. D

13. The gamete generation gives rise to gametes by:

    A. mitosis
    B. meiosis
    C. syngamy
    D. reduction division

    Ans. A

14. The sporophyte generation produces spores immediately following:

    A. mitosis
    B. meiosis
    C. fertilization
    D. gametangia formation

    Ans. B

15. When the sperm fertilizes the egg it results in the formation of:

    A. multicellular embryo
    B. diploid gametangium
    C. diploid sporangium
    D. diploid zygote

    Ans. D

16. The embryo of a plant matures into:

    A. a spore
    B. a rhizoid
    C. a sporophyte plant
    D. a gametophyte plant

    Ans. C

17. Each plant spore is capable of growing into:

    A. a multicellular sporophyte plant
    B. a multicellular gametophyte plant
    C. a gametangia
    D. a fertilized egg

    Ans. B

18. Xylem is to phloem as:

    A. conduction of dissolved food is to conduction of water and minerals
    B. lack of support is to support
    C. conduction of water and minerals is to conduction of dissolved food
    D. conduction of dissolved organic macromolecules is to conduction of inorganic micromolecules

    Ans. C

19. Ferns are classified as:

    A. seedless vascular plants that reproduce by spores
    B. vascular plants that reproduce by forming flowers
    C. seedless nonvascular plants that reproduce by asexual reproduction
    D. seedless nonvascular plants that reproduce by meiosis

    Ans. A

20. A basic difference between gymnosperms and flowering plants is:

    A. gymnosperms produce seeds that are enclosed within a fruit, while flowering plants produce seeds borne naked
    B. gymnosperms produce flowers that have floral parts arranged in threes, or multiples of threes, while flowering plants have floral parts arranged in fours or in multiples of four
    C. gymnosperms produce only cones, while flowering plants produce both flowers and cones
    D. gymnosperms produce seeds borne naked, while flowering plants produce seeds enclosed within a fruit

    Ans. D

21. The plants classified as bryophytes include:

    A. mosses, liverworts, and hornworts
    B. mosses, ferns, and whisk ferns
    C. mosses, club mosses, and liverworts
    D. mosses, horsetails, and ferns

    Ans. A

22. Mosses may represent an evolutionary sideline, evolved from ancestral green algae, or may have evolved from:

    A. photosynthetic bacteria
    B. nonphotosynthetic protozoans that underwent evolutionary change due to selection for photosynthesis and multicellularity
    C. vascular plants, that evolved by becoming simpler and losing their vascular tissue
    D. ancestral coral, that were gradually selected for terrestrial existence and photosynthesis

    Ans. C

23. The main advancement exhibited by ferns and their allies over mosses and other byrophytes is:

    A. presence of leaves
    B. presence of specialized vascular tissue
    C. presence of gametophyte generation and sporophyte generation
    D. presence of seeds

    Ans. B

24. Spore production in ferns usually occurs on:

    A. the stems that develop sporangia
    B. the rhizomes that develop sporangia
    C. the gametophytes that develop sporangia
    D. the fronds that develop sporangia

    Ans. D

25. In a fern, what is the name of a tiny, leaflike structure, that is often heart-shaped, that grows flat against the ground?

    A. gametophyte
    B. sporophyte
    C. rhizome
    D. diploid zygote

    Ans. A

26. Ferns are considered more advanced than mosses because they possess vascular tissues, but they have retained a primitive fertilization technique, which is:

    A. the use of wind as a transport medium
    B. the use of water as a transport medium
    C. the use of flowers which do not have petals, thus they are very small
    D. the use of ants as pollinators, as opposed to bees and butterflies

    Ans. B

27. The sporophyte generation in the fern is dominant for two reasons

    A. it is larger and produces sporangia that are very large and colorful
    B. it persists for a longer period of time and its sporangia are very colorful and flower-like
    C. it is larger than the gametophyte and it persists for an extended period of time
    D. it is larger and its reproductive functions are essential to the life cycle, whereas the gametophyte's reproductive functions are not essential

    Ans. C

28. A plant which produces only one type of spore is to a plant which produces two different types of spores, as:

    A. isospory is to disporous
    B. monospory is to bisporous
    C. microspory is to macrosporous
    D. homospory is to heterosporous

    Ans. D

29. Male gamete is to female gamete as:

A. isogamete is to heterogamete
B. microspore is to megaspore
C. pollen is to seed
D. microgametophyte is to megagametophyte

Ans. B

30. There are two groups of seed plants, which are referred to as:

A. gymnosperms and angiosperms
B. pines and flowering plants
C. monocots and dicots
D. herbaceous, nonwoody plants and woody plants

Ans. A

31. Seeds that are totally exposed or borne on scales of cones are to seeds that are enclosed in a vessel or within a fruit, as:

A. anthophyte is to gymnosperm
B. angiosperm is to anthophyte
C. gymnosperm is to angiosperm
D. pine is to ginkgo

Ans. C

32. A major advancement in the conifer life cycle, regarding reproduction is:

A. insects have replaced water as the medium for transporting the sperm
B. insects have replaced air as the medium for transporting pollen
C. fertilization occurs throughout the year as opposed to annual fertilization
D. water as a transport medium for sperm has been replaced by air currents which transport pollen

Ans. D

33. The fertilization process in flowering plants, is referred to as:

A. double fertilization
B. triple fertilization
C. polyploidal fertilization
D. sporophyte fertilization

Ans. A

Plant Life

34. Zygote is to endosperm as:

    A. sperm fertilizes egg is to sperm fuses with three cells in the female gametophyte
    B. sperm fertilizes egg is to sperm fuses with two cells in the female gametophyte
    C. monocot reproduction is to dicot reproduction
    D. two sperm fertilize the egg is to two sperm fertilize two cells in the female gametophyte

    Ans. B

35. An embryonic seed leaf in a flowering plant, is referred to as:

    A. synergistic body
    B. zygote
    C. endosperm
    D. cotyledon

    Ans. D

36. The main evolutionary advantage which flowering plants have over other plants, is their:

    A. ability to withstand variable temperature fluctuations
    B. ability to withstand variable amounts of moisture
    C. ability to attract animals for pollen dispersal, thereby assuring cross-fertilization
    D. ability to self-pollinate, thus assuring that every flower will be fertilized

    Ans. C

37. Compare alternation of generations in mosses and ferns, and explain which stage is dominant.

38. Explain why seeds are such a significant evolutionary development.

39. Compare the differences that exist between gymnosperms and angiosperms.

40. Name three groups of plants that are classified as:

    a. bryophytes
    b. pterophytes
    c. gymnosperms
    d. monocots
    e. dicots

Chapter 24

Animal Life--Invertebrates

1. Coral reefs represent:

   A. one of the most productive marine ecosystems and most common marine ecosystems
   B. one of the most productive marine ecosystems and most diverse
   C. the largest offshore marine ecosystem, and therefore the most productive
   D. the fastest growing marine ecosystem

   Ans. B

2. Coral reefs are not only beneficial to humans biologically, but they are also important economically, because they:

   A. form and maintain the foundation of many islands and they protect many costal shorelines against storms and erosion
   B. they are an important source of food in the tropics, because when they are ground, they are rich in phosphorous
   C. they are capable of absorbing and metabolizing toxic chemical wastes
   D. they are the economic backbone of tourism wherever they are found

   Ans. A

3. Coral reefs are made up of colonies of tiny animals related to:

   A. barnacles
   B. clams and oysters
   C. jellyfish
   D. sand dollars

   Ans. C

4. Corals live in a mutualistic symbiotic relationship with photosynthetic algae, which are referred to as:

   A. green algae
   B. kelp
   C. zoocarotids
   D. zooxanthellae

   Ans. D

5. Coral bleaching occurs when:

   A. coral is exposed to bleach pollutants
   B. coral is exposed to too much sunlight when water levels drop
   C. they lose their symbiotic algae
   D. the reef grows old, many of the younger corals grow over the older corals and reduce their food

   Ans. C

6. The precentage of distribution of vertebrate animals to invertebrate animals is approximately:

   A. 1% to 99%
   B. 5% to 95%
   C. 30% to 70%
   D. 50% to 50%

   Ans. B

7. All animals have some characteristics that they have in common with other animals, which are:

   A. they are multicellular eukaryotes, which are heterotrophic and most reproduce sexually
   B. they are multicellular prokaryotes which are heterotrophic and most reproduce sexually
   C. they are heterotrophs, which are motile, and most lower forms reproduce exclusively by asexual means
   D. a well-developed nervous system, alternation of generations, and all are organized into organ systems

   Ans. A

8. Animal dependency is to plant, as:

   A. heterotroph is to consumer
   B. autotroph is to heterotroph
   C. consumer is to producer
   D. photosynthesis is to consumer

   Ans. C

9. Sponges are to all other animals as:

   A. Protista is to Animalia
   B. soft body parts are to hard body parts
   C. parazoa is to vertebrata
   D. Parzoa is to Eumetazoa

   Ans. D

10. The three embryonic layers from which most animals develop are collectively called:

    A. derm layers
    B. germ layers
    C. zygotic layers
    D. formative layers

    Ans. B

11. No body cavity is to a true body cavity that is lined, as:

    A. pseudocoelom is to coelom
    B. pseudocoelom is to acoelom
    C. acoelom is to coelom
    D. acoelom is to eucoelom

    Ans. C

12. Blastopore and mouth is to blastopore and anus, as:

    A. protostome is to deuterostome
    B. chordate is to echinoderms
    C. mollusk is to arthropod
    D. deuterostome is to protostome

    Ans. A

13. Asexual reproduction is to sexual reproduction as:

    A. bud fragment is to hermaphroditic
    B. binary fission is to bud fragment
    C. fertilization is to zygote
    D. high diversity is to low diversity

    Ans. A

14. In Cniderians (hydras and jellyfish) that exhibit alternation of generation, the asexual stage is to the sexual stage as:

    A. medusa form is to planula form
    B. polyp form is to medusa form
    C. medusa form is to polyp form
    D. polyp form is to planula form

    Ans. B

15. The ciliated larval form of a cnidarian is produced by the:

    A. polyp
    B. planula
    C. medusa
    D. chidoblast

    Ans. C

16. Tentacles that bear cnidocytes (stinging cells) are to tentacles that do not bear enidocytes (they have adhesive glue cells), as:

    A. Porifera is to Cnidaria
    B. Cnideria is to Platyhelminthes
    C. Ctenophora is to Nemertea
    D. Cnidaria is to Ctenophora

    Ans. D

17. Cephalization is an important evolutionary adaptation which is first seen in the flatworms. Cephalization refers to:

    A. the development of organ systems
    B. the development of a head
    C. the development of a heart
    D. the development of a body cavity

    Ans. B

18. Flatworms that live in freshwater have an osmotic problem, because freshwater is hypotonic to their tissues. To maintain proper osmotic balance excess water must be discharged which is the function of the:

    A. protonephridia
    B. kidneys
    C. chidoblasts
    D. Malpighian tubules

    Ans. A

19. Carnivore is to parasite as:

    A. fluke is to tapeworm
    B. rotifer is to planaria
    C. ascaris is to planaria
    D. planaria is to fluke

    Ans. D

20. Tapeworms are thought of as the most degenerate type of flatworm, because they are lacking:

    A. sexual reproductive organ
    B. a nerve cord
    C. digestive organs and a mouth
    D. a head

    Ans. C

21. Two important evolutionary developments in ribbon worms are:

    A. a tube-within-tube body plan, and separate circulatory system
    B. a circulatory system, and eyes that can focus images as opposed to simply detecting light
    C. the sexes are separated or not found within the same individual, and they possess an exoskeleton
    D. proboscis that secretes mucous and is equipped with a poison gland, eyes that can focus images as opposed to simply detecting light

    Ans. A

22. Nematodes are the earliest animals known to have evolved a body cavity which is known as:

    A. acoelom
    B. pseudocoelom
    C. endocoelom
    D. coelom

    Ans. B

23. Ascaris worms which are common parasites in humans are passed from one individual to another in the following manner:

    A. human eats an infected fish that is not properly cooked
    B. human is bitten by a mosquito that is infected with the worm's eggs
    C. human feces contain eggs, and where sanitation is poor they find their way into the soil
    D. human eats pork that has been infected due to poor sanitation

    Ans. C

24. Small aquatic animals that are known as "wheel animals" belong to the phylum:

    A. Nematoda
    B. Gastropoda
    C. Mollusca
    D. Rotifera

    Ans. D

Animal Life--Invertebrates

25. Snails, clams, oysters, squid, octopods, and slugs all belong to the phylum:

    A. Cephalopoda
    B. Mollusca
    C. Gastropoda
    D. Parapoda

    Ans. B

26. Annelids have evolved segmentation which is very important from an evolutionary perspective because it provides the opportunity for:

    A. specialization of body regions
    B. development of a separate excretory system and digestive tract
    C. development of greater locomotion and more complex nervous system
    D. development of separate respiratory and circulatory systems

    Ans. A

27. Sandworms and tubeworms are to earthworms as:

    A. Annelida is to Polychaeta
    B. Ollgochaeta is to Annelida
    C. Polychaeta is to Oligochaeta
    D. hydrostatic skeleton is to endoskeleton

    Ans. C

28. Earthworms are important because they:

    A. consume their weight in harmful pathogenic bacteria
    B. are vital to the formation and maintenance of fertile soil
    C. consume and destroy many harmful insect eggs that are present in the soil
    D. produce the anticoagulant Hirudin, which is used to prevent blood from clotting

    Ans. B

29. Horseshoe crabs, spiders, ticks, scorpions, lobsters, barnacles, and centipedes belong to the phylum:

    A. Chelecerata
    B. Arachnida
    C. Crustacea
    D. Arthropods

    Ans. D

30. The process of shedding an old exoskeleton and growing another is known as:

A. exocytosis
B. metamorphosis
C. molting
D. exoskeleton detachment

Ans. C

31. Terrestrial insects and spiders have specialized respiratory systems that consist of many fine branching air tubes or platelike structures which are respectively called:

A. tracheae and book lungs
B. gill lungs and ostia
C. merastoma and tracheated hexapoda
D. malpighian tubules and spiracles

Ans. A

32. The circulatory system of an arthropod is classified as:

A. a closed circulatory system
B. a tube-within-tube circulatory system
C. a branching circulatory system
D. an open circulatory system

Ans. D

33. Several development stages or nymphal stages are to four distinct stages in the life cycle (egg, larva, pupa, and adult), as:

A. partial metamorphosis is to complete metamorphosis
B. gradual metamorphosis is to complete metamorphosis
C. articulated development is to disarticulated development
D. indirect development is to direct development

Ans. B

34. One of the most important evolutionary adaptations of insects is their ability to:

A. see images
B. sting or bite
C. fly
D. smell and taste

Ans. C

Animal Life--Invertebrates

35. Sea urchins, sand dollars, sea stars, brittle stars, sea lillies, and sea cucumbers all belong to the phylum:

    A. Uniramia
    B. Crustacea
    C. Chelicerata
    D. Echinodermata

    Ans. D

36. What type of symmetry do adult sea stars, sea urchins, sand dollars, and sea cucumbers exhibit?

    A. pentaradial symmetry
    B. bilateral symmetry
    C. radial symmetry
    D. asymmetry

    Ans. A

37. Differentiate between insects and spiders.

38. Differentiate between the following phyla, using distinguishing characteristics.
    a. phylum Echinodermata
    b. phylum Annelida
    c. phylum Arthropoda
    d. phylum Mollusca

39. Draw a hypothetical ancestral tree of the invertebrate animals, using generally accepted evolutionary principles.

40. Compare and the contrast the characteristics of acoelomates, pseudocoelomates, and coelomates.

Chapter 25

Animal Life: Chordates

1. A sudden mass extinction occurred about 65 million years ago, which would correspond to which geological period?

   A. Tertiary
   B. Jurassic
   C. Crustaceous
   D. Permian

   Ans. C

2. The mass extinction that occurred approximately 65 million years ago was possibly caused by one or more asteroids or large comets, which hit the Earth, drastically altering global ecology. The asteroids affected the Earth and its plants and animals by:

   A. causing massive tidal waves which resulted in mass extinction
   B. causing massive earthquakes the resulted in mass extinction
   C. causing massive shock waves that resulted in mass extinction
   D. causing massive clouds of dust that locked out the sunlight for several months or possibly years

   Ans. D

3. What element is extremely rare in the Earth's crust but is comparatively abundant in extraterrestrial objects, which gives scientists a clue to the extraterrestrial hypothesis?

   A. radon
   B. iridium
   C. osmium
   D. radium

   Ans. B

4. Two rather large craters have been found in Mexico and Iowa, that are estimated to be 65.7 million years old, which coincides with the estimated time of the mass extinction. The technique investigators used to determine the age of the craters was:

   A. argon dating
   B. carbon dating
   C. uranium dating
   D. boron dating

   Ans. A

5. The subphyla Urochordata, Cephalochordata, and Vertebrata are part of the phylum:

   A. Euchordata
   B. Chordata
   C. Animalia
   D. Protochordata

   Ans. B

6. Tunicates (sea squirts) are to lancelets (fishlike animals) as

   A. Cephalochordata is to Vertebrata
   B. Urochordata is to Vertebrata
   C. Urochordata is to Cephalochordata
   D. Agnatha is to chondrichthyes

   Ans. C

7. A dorsal longitudinal rod for support of the body is to a tail that extends beyond the anus, as:

   A. nerve cord is to prenatal tail
   B. backbone is to postnatal tail
   C. endoskeleton is to anal tail
   D. notochord is to postanal tail

   Ans. D

8. A hollow cord located on the back surface is to perforations through the wall of the pharynx, as:

   A. ventral tubular nerve cord is to pharyngeal gill grooves
   B. dorsal tubular nerve cord is to pharyngeal gill grooves
   C. dorsal tubular nerve cord is to inner and middle ear
   D. tubular nerve cord is to functional gill slits

   Ans. B

9. Sea squirts which are often mistaken for sponges or cnidarians, have the following characteristics:

   A. filter feeders, reproduce asexually by budding, and are usually hermaphroditic
   B. have tentacles for feeding, reproduce asexually by binary fission, and are usually hermaphroditic
   C. have a spongoceol like a sponge, but have tentacles and cnidocils (stinging cells) like cnidarians
   D. are sessile (stationary) like a sponge, but have a polyp and medusa stage like a cnidarian

   Ans. A

10. Lancelets resemble fishes, but their body plan is far simpler. Lancelets lack:

   A. muscles, a well defined brain, and digestive system
   B. sexual reproductive organs (they reproduce asexually), a circulatory system, and an anus
   C. jaws, paired fins, and a heart
   D. sense organs, a well defined brain, and striated muscles

   Ans. C

11. Most vertebrates are distinguished from other chordates by having:

   A. functional notochord
   B. functional eyes
   C. a vertebral column
   D. a heart and open circulatory system

   Ans. C

12. Segments in the vertebral column are to the anatomical feature located anterior to the vertebral column, as:

   A. backbones are to brain
   B. nerve cord is to brain
   C. notochord is to cephalon
   D. vertebrae is to cranium

   Ans. D

13. Living skeleton is to nonliving skeleton, as:

   A. biotic skeleton is to abiotic skeleton
   B. endoskeleton is to exoskeleton
   C. exoskeleton is to hydrostatic skeleton
   D. exoskeleton is to endoskeleton

   Ans. B

14. The type of vertebrate circulatory system is to the location of the vertebrate heart as:

    A. open circulatory system is to medial location
    B. closed circulatory system is to dorsal location
    C. closed circulatory system is to ventral location
    D. dynamic circulatory system is to medial location

    Ans. C

15. Jawless fishes are to fishes with a cartilage skeleton as:

    A. lampreys are to sharks
    B. lancelets are to tunicates
    C. lancelets are to lampreys
    D. lampreys are to hagfishes

    Ans. A

16. Amphibians are to reptiles, as:

    A. salamanders and lizards are to snakes and turtles
    B. frogs and toads are to salamanders and alligators
    C. snakes and turtles are to toads and frogs
    D. frogs and salamanders are to alligators and lizards

    Ans. D

17. Sharks have toothlike scales that are referred to as:

    A. cosmoid scales
    B. placoid scales
    C. ganold scales
    D. cycloid scales

    Ans. B

18. Close relatives of sharks that are flattened, that live on the bottom and feed on mussels and clams are referred to as:

    A. rays and suckers
    B. flounders and suckers
    C. rays and skates
    D. flukes and porgys

    Ans. B

19. A lateral protective flap that extends posteriorly from the head over the gills is to fishes that lay eggs, as:

    A. fin ray is to viviparous
    B. lateral ray is to ovoviviparous
    C. operculum is to ovoviviparous
    D. operculum is to oviparous

    Ans. D

20. The ancestors of the ray-finned fishes are thought to have had lungs. These lungs later became modified as:

    A. a swim bladder
    B. a pair of internal gills
    C. a blood filtering mechanism
    D. a crop

    Ans. A

21. Amphibians most likely evolved from some type of ancestral fishes referred to as:

    A. bony-finned fishes
    B. scale-finned fishes
    C. lobe-finned fishes
    D. ray-finned fishes

    Ans. C

22. First successful land vertebrates are to the evolutionary link of two modern forms and one early form, as:

    A. tetrapods are to frogs, toads, and reptiles
    B. labyrinthodonts are to frogs, salamanders, and cotylosaurs
    C. cotylosaurs are to frogs, salamanders, and earliest reptiles
    D. cotylosaurs are to frogs, caecilians, and labyrinthodonts

    Ans. B

23. Frog and toad tadpoles are to the amphibian heart, as:

    A. nymphs are to two chambers
    B. larvae are to two chambers
    C. larvae are to three chambers
    D. fry are to four chambers

    Ans. C

24. An amphibian circulatory system consists of how many circuits of blood vessels?

    A. single circuit in which oxygen-rich blood mixes with oxygen-poor blood
    B. two circuits in which there is no mixing of oxygen-rich and oxygen-poor blood
    C. double circuit which keeps oxygen-rich and oxygen-poor blood partially separated
    D. triple circuit in which some oxygen-rich and oxygen-poor blood is totally mixed, some partially mixed, and the remainder is separated

    Ans. C

25. Amphibian skin is to reptilian skin, as:

    A. secretes mucous and poisonous substances. and is slippery are to hard, dry, horny scales that lack skin secretions
    B. secretes mucous and poisonous substances, and is slippery are to horny scales that secrete mucus and poisonous substances, and is slippery
    C. naked skin that lacks scales and mucus glands, but is slippery due to water are to horny, dry, scales that when wet become slippery
    D. skin with small underdeveloped scales that secrete mucus, and is slippery are to skin with well developed scales and underdeveloped mucus glands

    Ans. A

26. All reptiles reproduce sexually by:

    A. external fertilization in which eggs are fertilized in water
    B. internal fertilization
    C. external fertilization in which eggs are fertilized in a moist terrestrial environment
    D. internal fertilization in which the male deposits a spermatophore outside the body, which is subsequently drawn into the female's body and spermatzoa are released

    Ans. B

27. The membrane that develops and surrounds the embryo, is referred to as:

    A. chorion
    B. amnion
    C. leathery shell
    D. ectoderm

    Ans. B

28. Organisms that lack metabolic mechanisms for regulating body temperature are to organisms that have evolved metabolic mechanisms for maintaining a constant body temperature, as:

    A. endothermic is to homeothermic
    B. endothermic is to poikilothermic
    C. ectothermic is to endothermic
    D. ectothermic is to poikilothermic

    Ans. C

29. Mammals evolved from a group of reptiles called:

    A. plethoconts
    B. euryceaids
    C. bufapsids
    D. therapsids

    Ans. D

30. Early mammals were specialized in being inconspicuous, because they were adapted to:

    A. living fossilorially and diurnally
    B. living arboreally and nocturnally
    C. living hypogeanally and nocturnally
    D. living aquatically and nocturnally

    Ans. B

31. The mammals were able to expand into new niches and habitats, as a particular class died out. Which organisms made it possible for mammals to expand?

    A. reptiles
    B. birds
    C. amphibians
    D. fishes

    Ans. A

32. Egg-laying mammals are to pouched mammals as:

    A. prosimians are to hominoids
    B. prosimians are to monotremes
    C. monotremes are to marsupials
    D. marsupials are to therapsids

    Ans. C

33. The earliest primates probably evolved from the tree dwelling mammals, which were:

    A. orangutans
    B. chimpanzees
    C. monkeys
    D. shrewlike mammals

    Ans. D

34. Lemurs, lorises, and tarsiers are to monkeys, apes, and humans, as:

    A. prosimians are to primates
    B. prosimians are to autropoids
    C. autropoids are to hominoids
    D. primates are to hominoids

    Ans. B

35. Human evolution occurred in which sequence of hominoids?

    A. *Homo habilis*, *Homo erectus*, and *Homo sapiens*
    B. *Homo erectus*, *Homo habilis*, and *Homo sapiens*
    C. *Australopithecus afarensis*, and *Homo sapiens*
    D. *Australopithecus africansus*, *Homo sapiens* and *Homo habilis*

    Ans. A

36. The earliest *Homo sapiens*, were referred to as:

    A. inuits
    B. Peking
    C. Neanderthals
    D. Cro-Magnon

    Ans. C

37. Explain why vertebrates survived so successfully when they became terrestrial.

38. Compare and explain why reptiles are more advanced evolutionarily than amphibians.

39. Discuss and explain the probable cause for the extinction of the dinosaurs, and why mammals were able to profit from their extinction.

40. Why are birds able to inhabit the polar regions but reptiles do not?

Chapter 26

Plant Structure

1. Approximately what percentage of all prescription medicines are derived from plants?

   A. 10%
   B. 25%
   C. 50%
   D. 75%

   Ans. B

2. Many of the plant-produced chemicals with medicinal properties contain organic compounds that contain nitrogen. This group of plant-produced chemicals are referred to as:

   A. carotenoids
   B. xanthoids
   C. phenols
   D. alkaloids

   Ans. D

3. The study of the traditional uses of plants by indigenous people, is referred to as:

   A. ethnobotany
   B. pharmaceutical botany
   C. medicinal botany
   D. ecobotany

   Ans. A

4. Only about 5,000 of the 350,000 species of flowering plants have been investigated for their medicinal value. Unfortunately, we may never have the opportunity to discover their value, because they are threatened:

   A. due to uncontrolled herbivorous insects
   B. due to changing climates
   C. due to loss of habitat
   D. due to indiscriminate use of herbicides

   Ans. C

Plant Structure

5. "Soft-stemmed" plants are to "hard-stemmed" plants as:

   A. nonvascular plants are to vascular plants
   B. bryophytes are to tracheophytes
   C. herbaceous plants are to woody plants
   D. grasses are to forbes

   Ans. C

6. A one-year life cycle is to a two-year life cycle, as:

   A. monoennial is to diennial
   B. annual is to biannual
   C. yearly is to biyearly
   D. ennial is to diennial

   Ans. B

7. Plants that normally live for an extended period of time are referred to as:

   A. polyennials
   B. multiennials
   C. perpetuals
   D. perennials

   Ans. D

8. The plant is organized into two systems, which are referred to as:

   A. root system and shoot system
   B. nonreproductive system and reproductive system
   C. nonphotosynthetic system and photosynthetic system
   D. asexual reproductive system and sexual reproductive system

   Ans. A

9. Two types of roots occur in plants, which are referred to as:

   A. primary and secondary roots
   B. fibrous roots and taproots
   C. meristematic roots and mature roots
   D. root hairs and differentiated roots

   Ans. B

10. Location of leaf attachment is to an embryonic shoot at the distal end of a stem, as:

    A. node is to terminal bud
    B. internode is to meristem
    C. leaf bud is to meristem
    D. promordial bud is to axil bud

    Ans. A

11. Modified leaves are to loosely arranged cells that allow diffusion to occur in the stem, as:

    A. stomata are to epidermis
    B. lenticels are to stomata
    C. leaf primordia are to stomata
    D. bud scales are to lenticels

    Ans. D

12. The part of the bud which acts as a stalk for attachment to the stem, is referred to as:

    A. vascular trace
    B. lamina
    C. petiole
    D. bundle sheath

    Ans. C

13. The flat, or broad portion of the leaf, is referred to as:

    A. sheath
    B. blade
    C. leaflet
    D. axil

    Ans. B

14. Undivided leaves are to leaves that are divided into smaller leaves, as:

    A. monastic leaves are to polyastic leaves
    B. simple leaves are to complex leaves
    C. simple leaves are to differential leaves
    D. single leaves are to compound leaves

    Ans. D

Plant Structure

15. The leaf arrangement in which one leaf originates at the point of origin on the stem is referred to as:

    A. alternate
    B. opposite
    C. whorled
    D. distinctive

    Ans. A

16. The leaf arrangement in which three leaves originate at the same point of origin on the stem, is referred to as:

    A. alternate
    B. opposite
    C. whorled
    D. distinct differentiated

    Ans. C

17. Parallel venation is to netted venation as:

    A. nonvascular is to vascular
    B. monocot is to dicot
    C. gymnosperm is to angiosperm
    D. dicot is to monocot

    Ans. B

18. In plants a group of cells that form a structural and functional unit is referred to as:

    A. tissue
    B. organ
    C. system
    D. organism

    Ans. A

19. The ground tissue system is composed of three tissues, each of which have different functions. They include the following functions:

    A. photosynthesis, reproduction, and transport
    B. photosynthesis, reproduction, and absorption
    C. growth, reproduction, and energy utilization
    D. photosynthesis, storage, and support

    Ans. D

20. Conduction and support are to the covering of the plant as:

    A. pith is to tegument
    B. vascular tissue system is to dermal tissue system
    C. phloem is to epithelium
    D. xylem is to ectoderm

    Ans. B

21. A growing cell secretes a thin partition which stretches and expands as the cell increases in size. It is referred to as:

    A. initial plasma wall
    B. temporary cell wall
    C. primary cell wall
    D. tertiary cell wall

    Ans. C

22. After the cell stops growing, it sometimes secretes a thick partition, which is referred to as:

    A. permanent cell wall
    B. primary cell wall
    C. tertiary cell wall
    D. secondary cell wall

    Ans. D

23. Rather simple tissues that perform functions such as photosynthesis, storage, and secretion, are classified as:

    A. parenchyma
    B. meristematic
    C. collenchyma
    D. periderm

    Ans. A

24. A rather simple tissue, that is extremely flexible support tissue that provides much of the support in soft, nonwoody plant organs, is referred to as:

    A. parenchyma
    B. sclerenchyma
    C. collenchyma
    D. cortex

    Ans. C

Plant Structure

25. A simple tissue specialized for structural support, in which at maturity the cells are often dead. The secondary cell walls of these cells become strong and hard due to extreme thickening. This tissue is referred to as:

    A. collenchyma
    B. sclerenchyma
    C. parenchyma
    D. periderm

    Ans. B

26. Xylem is to phloem, as:

    A. conducts dissolved sugars is to conducts amino acids and water
    B. conducts water and dissolved minerals is to conducts dissolved salts
    C. conducts dissolved organic compounds is to conducts inorganic compounds
    D. conducts water and dissolved minerals is to conducts dissolved sugar

    Ans. D

27. Xylem is to phloem as:

    A. tracheids and companion cells are to vessel elements and sieve-tubes
    B. rays and vessel elements are to sieve-tubes and pericycle
    C. tracheids and vessel elements are to sieve-tubes and companion cells
    D. vessel elements and stele cells are to companion cells and cortex cells

    Ans. C

28. Woody plants initially produce epidermis, but as the plant increases in diameter due to the production of additional woody tissues, the epidermis is replaced by a tissue referred to as:

    A. periderm
    B. collenchyma
    C. dermis
    D. stratified squamous epithelium

    Ans. A

29. Inward leaf diffusion is to outward leaf diffusion, as:

    A. carbon dioxide and water is to oxygen
    B. oxygen and carbon dioxide is to water
    C. carbon dioxide is to water and oxygen
    D. oxygen and water is to carbon dioxide

    Ans. C

30. Epidermal hairs which are involved in the absorption of water and minerals, or removal of excess salt and reflection of light, are referred to as:

    A. cuticles
    B. trichomes
    C. cork
    D. granulors

    Ans. B

31. Growth is a complex phenomenon involving three different processes, which occur in the following sequence:

    A. cell elongation, cell division, and cell differentiation
    B. primary growth, secondary growth, and cell differentiation
    C. cell differentiation, cell specialization, and cell maturity
    D. cell division, cell elongation, and cell differentiation

    Ans. D

32. One difference between the growth patterns of animals and plants is:

    A. animal growth occurs in specific regions, while plant growth is throughout the entire plant
    B. animal growth occurs throughout the entire localized into specific areas
    C. animal growth is restricted to mitosis, combination of mitosis and meiosis
    D. animal growth is initially mitotic, but as replaces mitosis, while plants are restriced life cycles

    Ans. B

33. Plant growth occurs in areas, referred to as:

    A. meristems
    B. growth centers
    C. maturation centers
    D. primordial growth regions

    Ans. A

34. Increases in the length of the plant is to increase in the girth, as:

    A. net growth is to total growth
    B. herbaceous growth is to woody growth
    C. primary growth is to secondary growth
    D. apical growth is to perimeter growth

    Ans. C

Plant Structure

35. Region of cell division that results in longitudinal growth is to region of cell division that results in increasing the girth, as:

    A. vascular cambium is to cork cambium
    B. proximal meristem is to lateral meristem
    C. distal meristem is to medial meristem
    D. apical meristem is to lateral meristem

    Ans. D

36. Embryonic leaves are to embryonic buds, as:

    A. meristems are to cambia
    B. leaf primordia are to bud primordia
    C. photosynthetic meristems are to vascular meristems
    D. leaf mitotic centers are to bud mitotic centers

    Ans. B

37. Differentiate and explain the functions of xylem and phloem.

38. Differentiate and explain the functions of the cuticle and guard cells.

39. Explain why meristematic tissue is so important to plants.

40. Why are collenchyma and sclerenchyma common in shoots but uncommon in roots?

Chapter 27

Leaves

1. Most leaves are covered by:

   A. parenchyma and sclerenchyma cells
   B. lateral and medial cells
   C. upper epidermis and lower epidermis
   D. upper epithelium and lower epithelium

   Ans. C

2. Most of the cells that comprise the outer covering of the leaf are living parenchyma cells that generally:

   A. contain numerous chloroplasts and possess cytoplasm that is relatively transparent
   B. lack chloroplasts and are relatively transparent
   C. contain numerous chloroplasts and possess cytoplasm that is relatively opaque
   D. secrete alkaloids to protect the leaf externally and produce phenols internally to protect the interior

   Ans. B

3. Many of the cells that comprise the outer covering of the leaf are covered with hairs which are referred to as:

   A. dermosomes
   B. cristae
   C. guard cells
   D. trichomes

   Ans. D

4. A noncellular, waxy layer produced by the cells that comprise the outer covering of the leaf, is referred to as:

   A. cuticle
   B. mesophyll
   C. sulin
   D. sebaceous

   Ans. A

Leaves

5. Generally, the cells that comprise the outer covering of the leaf, and the noncellular waxy layer secreted by these cells, differ regarding cell size and thickness:

   A. upper layer of cells are smaller and the noncellular secretion is thinner when compared to the lower layer
   B. upper layer of cells are larger and the noncellular secretion is thicker when compared to the lower layer
   C. upper layer of cells are larger when compared to the lower layer of cells but the noncellular secretion is thicker when compared to the lower layer
   D. lower layer of cells are larger when compared to the upper layer of cells, but the noncellular secretion is thicker on the upper layer of cells

   Ans. B

6. Guard cells are the only cells that comprise the outer covering of the leaf that possess:

   A. mitochondria
   B. nucleus
   C. chloroplasts
   D. vacuoles

   Ans. C

7. Guard cells regulate the size of the:

   A. stomata
   B. cell thickness
   C. amount of noncellular secretion produced
   D. leaf surface to volume ratio

   Ans. A

8. Most guard cells are located on the lower surface because it is an adaptation that:

   A. reduces the loss of oxygen
   B. increases the absorption of carbon dioxide
   C. reduces the entry of microbes into the leaf
   D. reduces water loss

   Ans. D

9. The photosynthetic parenchyma tissue that comprises the internal portion of the leaf, is referred to as:

   A. mesoderm
   B. mesophyll
   C. metaphyll
   D. mesenchyma

   Ans. B

10. The upper layer of photosynthetic parenchyma cells that are toward the upper layer of cells that comprise the outer covering of the leaf, are referred to as:

    A. polarphyll
    B. spongy
    C. palisade
    D. stroma

    Ans. C

11. The lower layer of photosynthetic parenchyma cells that are located in the lower portion of the leaf, are referred to as:

    A. palisade
    B. polarphyll
    C. grana
    D. spongy

    Ans. D

12. The primary site for photosynthesis is to secondary site of photosynthesis, as:

    A. palisade is to mesenchyma
    B. palisade is to spongy
    C. spongy is to mesophyll
    D. stroma is to grana

    Ans. B

13. Primary function of diffusion is to secrete stinging irritants, as:

    A. spongy is to trichomes
    B. mesophyll is to guard cells
    C. palisade is to cnidocils
    D. mesochyma is to cuticle

    Ans. A

14. Leaves that are exposed to direct sunlight are:

    A. thinner that leaves that are exposed to in direct sunlight
    B. the same thickness are leaves that are exposed to indirect sunlight
    C. thicker than leaves that are exposed to indirect sunlight
    D. lighter in color than leaves that are exposed to indirect sunlight

    Ans. C

Leaves

15. The veins of a leaf extend through the:

    A. metaderm
    B. mesoderm
    C. mesenchyma
    D. mesophyll

    Ans. D

16. Upper side of leaf is to lower side of leaf, as:

    A. phloem is to xylem
    B. xylem is to phloem
    C. spongy is to palisade
    D. guard cell is to xylem

    Ans. B

17. Veins are usually surrounded by one or more layers of cells, which are referred to as:

    A. bundle subsidiaries
    B. bundle components
    C. bundle sheath
    D. nonvascular support tissue

    Ans. C

18. Monocot is to dicot, as:

    A. narrow leaves with parallel venation is to broad leaves with netted venation
    B. broad leaves with parallel venation is to narrow leaves with netted venation
    C. one or two cotyledons is to two or three cotyledons
    D. narrow leaves with a distinct petiole is to broad leaves lacking a petiole

    Ans. A

19. Monocot is to dicot as:

    A. palms, corn, and roses are to beans, oaks, and grasses
    B. tulips, orchids, and bananas are to apples, cherries, and snapdragons
    C. wheat, corn, and elms are to grasses, maples, and peppers
    D. maples, oaks, and palms are to wheat, rice, and corn

    Ans. B

20. Guard cells of both monocots and dicots may be associated with special epidermal cells, which are referred to as:

    A. subcutaneous cells
    B. subaceous cells
    C. subdermal cells
    D. subsidiary cells

    Ans. D

21. The primary function of leaves is:

    A. photosynthesis
    B. cellular respiration
    C. transpiration
    D. collect radiant energy and convert it into inorganic molecules

    Ans. A

22. Sugar that is a product of the leaves can follow two biochemical pathways, which are:

    A. release energy and cellular respiration
    B. convert sugar into glucose and starch
    C. release chemical energy and convert glucose into starch and cellulose
    D. release chemical energy and convert starch and cellulose into glucose

    Ans. C

23. Photosynthetic reactant is to photosynthetic product, as:

    A. oxygen is to carbon dioxide
    B. water is to carbon dioxide
    C. glucose is to oxygen
    D. carbon dioxide is to oxygen

    Ans. D

24. Air photosynthetic reactant is to soil photosynthetic reactant as:

    A. oxygen is to water
    B. carbon dioxide is to water
    C. water is to dissolved minerals
    D. carbon dioxide is to chlorophyll

    Ans. B

Leaves

25. The leaf, which is structurally weak because of large amounts of air space, receives structural support from:

    A. epidermal layers
    B. palisade layer
    C. bundle sheaths
    D. endodermis and ectodermis

    Ans. C

26. Turgid cells are to flaccid cells as:

    A. stomata open are to stomata closed
    B. nonphotosynthetic reactions are to photosynthetic reactions
    C. stomata closed are to stomata open
    D. nontranspiration reactions are to transpiration reactions

    Ans. A

27. Stomata open in response to:

    A. high levels of potassium ions and high levels of carbon dioxide
    B. high levels of calcium ions and low levels of oxygen
    C. low levels of potassium ions and low levels of carbon dioxide
    D. high levels of potassium ions and low levels of carbon dioxide

    Ans. D

28. Biological rhythms that follow an approximate 24-hour cycle are known as:

    A. homeostatic rhythms
    B. physiological rhythms
    C. circadian rhythms
    D. nocturnal - diurnal rhythms

    Ans. C

29. Stomatal opening is most pronounced in:

    A. red light
    B. blue light
    C. green-yellow light
    D. violet light

    Ans. B

30. Transpiration refers to:

    A. loss of oxygen from the aerial parts
    B. loss of carbon dioxide from aerial parts
    C. exchange of carbon dioxide and oxygen in aerial parts
    D. loss of water vapor from aerial parts

    Ans. D

31. Environmental factors which increase the rate of transpiration are:

    A. increases in light, temperature, and relative humidity
    B. decreases in light, temperature, and relative humidity
    C. increase in light, and decreases in temperature and relative humidity
    D. increases in precipitation, barometric pressure, and temperature

    Ans. A

32. Removed by transpiration is to retained by transpiration, as:

    A. oxygen is to water
    B. water is to minerals
    C. carbon dioxide is to water
    D. oxygen is to heat

    Ans. B

33. The loss of liquid water from the leaves of plants, is referred to as:

    A. transpiration
    B. evaporation
    C. photoevaporation
    D. guttation

    Ans. D

34. What is the term that applies to the process in which the leaves fall off as winter approaches?

    A. guttate
    B. capitate
    C. abscise
    D. transpire

    Ans. C

Leaves

35. Leaves fall as winter approaches because it is critical to:

    A. conserve oxygen and carbon dioxide
    B. reduce the loss of water
    C. increase the consumption of minerals that act as natural antifreeze
    D. increase transpiration

    Ans. B

36. Leaves that are modified for storage of water or food, are referred to as:

    A. bulbs
    B. tubers
    C. tendrils
    D. corms

    Ans. A

37. Explain how stomatal openings are related to photosynthesis and transpiration.

38. Explain the series of physiological changes that occur in guard cells during stomatal opening.

39. Why do deciduous plants in temperate climates lose their leaves in autumn, and in the tropics broad-leafed plants do not lose their leaves in the winter?

40. How are leaves modified to tolerate dry environments? Aquatic environments?

Chapter 28

Stems and Roots

1. The main causes of soil degradation and erosion are farming practices that are environmentally unsound, and:

   A. housing developments and road construction
   B. overgrazing by livestock and deforestation
   C. draining of wetlands and estuaries
   D. expansion of beaches and lake fronts

   Ans. B

2. The black or dark brown decomposed organic material that makes up the upper horizon of the soil, is referred to as:

   A. fertile soil
   B. agricultural soil
   C. tomus
   D. humus

   Ans. D

3. The three primary functions of stems, are:

   A. food storage, photosynthesis, and support
   B. vegetative propagation, food storage, and conduction
   C. conduction, support, and production of new stem tissue
   D. photosynthesis, conduction, and production of new stem tissue

   Ans. C

4. Apical meristems are to lateral meristems, as:

   A. primary growth is to secondary growth
   B. primary growth is to tertiary growth
   C. secondary growth is to primary growth
   D. girth is to longitudinal growth

   Ans. A

Stems and Roots

5. What is the only type of growth which herbaceous stems exhibit?

   A. tertiary growth
   B. secondary growth
   C. latitudinal growth
   D. primary growth

   Ans. D

6. Mitosis is restricted to areas which are referred to as:

   A. growth centers
   B. meristems
   C. areas of elongation and differentiation
   D. rays

   Ans. B

7. The arrangement of tissues in the stem varies between the two groups of flowering plants, which are referred to as:

   A. angiosperms and gymnosperms
   B. herbaceous and woody plants
   C. monocots and dicots
   D. nonvascular and vascular

   Ans. C

8. Sunflowers and maples are to corn and wheat, as:

   A. nonvascular plants are to vascular plants
   B. angiosperms are to gymnosperms
   C. woody plants are to herbaceous plants
   D. dicots are monocots

   Ans. D

9. The outermost layer of cells on a plant stem, are referred to as:

   A. epidermis
   B. ectoderm
   C. epithelial
   D. ectothelial

   Ans. A

10. In a sunflower stem, the multiple layer of cells which is located to the inside of the outermost layer of cells, is referred to as:

A. cambium
B. cortex
C. sclerenchyma
D. pith

Ans. B

11. In a sunflower stem, the vascular tissue is located and arranged in:

A. a continuous circle
B. concentric annual rings
C. continuous rays
D. patches arranged in a circle

Ans. B

12. Beginning on the outside and continuing inward, the correct sequence of tissues in a vascular bundle of a sunflower stem is:

A. xylem, vascular cambium, phloem, and phloem fiber cap
B. phloem fiber cap, phloem, xylem, and vascular cambium
C. phloem fiber cap, phloem, vascular cambium, and xylem
D. xylem, phloem, phloem fiber cap, and vascular cambium

Ans. C

13. The center of the dicot stem is occupied by a tissue, referred to as:

A. cambium
B. pith
C. xylem
D. ray

Ans. B

14. The vascular bundles of dicot stems alternate with a tissue classified as:

A. pith rays
B. vascular rays
C. phloem rays
D. xylem rays

Ans. A

Stems and Roots

15. In a vascular bundle of a monocot stem, the tissue located towards the inside is to the tissue located towards the outside, as:

    A. pith ray is to xylem ray
    B. phloem fiber cap is to xylem
    C. apical meristem is to vascular cambium
    D. xylem is to phloem

    Ans. D

16. A primary tissue in stems that provides protection is to a waxy layer that covers the primary tissue that provides protection, as:

    A. ectoderm is to suberin
    B. epidermis is to suberin
    C. epidermis is to cuticle
    D. epithelial is to Casparian strip

    Ans. C

17. Water and dissolved minerals are to dissolved sugar, as:

    A. phloem is to xylem
    B. xylem is to phloem
    C. xylem is to vascular bundle
    D. phloem fiber cap is to xylem

    Ans. B

18. The primary functions of roots are:

    A. absorption, conduction, cellular respiration, and exchange of ions
    B. anchorage, exchange of ions, conduction, and cellular respiration
    C. cellular respiration, anchorage, conduction, and exchange of metabolites
    D. anchorage, absorption, storage, and conduction

    Ans. D

19. The aerial roots of certain orchids function in:

    A. trapping insects
    B. collecting carbon dioxide and oxygen
    C. photosynthesis
    D. producing ammonia and nitrites

    Ans. C

20. Short-lived extensions of the epidermal layer in roots are to a protective layer that covers the apical meristem, as:

A. root hairs are to root cap
B. ectodermal cortex cells are to root extension
C. secondary roots are to phloem fiber cap
D. secondary roots are to dead cells that are sloughed off

Ans. A

21. The amount of surface area is increased and hence the absorptive capacity of roots increases, due to the presence of:

A. secondary roots
B. fibrous roots
C. root hairs
D. primary roots

Ans. C

22. Modified root epidermal cells are to secretion that would impede the absorption of water from the soil, as:

A. root hairs are to suberin
B. root hairs are to cuticle
C. secondary roots are to suberin
D. fibrous roots are to glycerol

Ans. B

23. The primary cells of the cortex of a dicot root are to inner layer of cortex cells of the dicot root, as:

A. parenchyma cells are to Casparian strip
B. pericycle is to endodermis
C. parenchyma cells are to pericycle
D. parenchyma cells are to endodermis

Ans. D

24. Cells which produce suberin are to the radial and transverse walls which contain suberin, as:

A. mesodermis is to cuticle
B. mesodermis is to vessel walls
C. endodermis is to Casparian strip
D. endothelium is to cell walls

Ans. C

Stems and Roots

25. The movement of water through the cortex of the root is to the principle function of the root cortex, as:

    A. intercellular space is to storage
    B. intracellular space is to synthesis of lipids
    C. intracellular space is to diffusion of oxygen and cellular respiration
    D. cytoplasmic flow is to diffusion of oxygen and cellular respiration

    Ans. A

26. The cellular layer of the root which is associated with meristematic activities is to the root structures formed as a result of meristematic activity, as:

    A. endothelium is to fibrous roots
    B. pericycle is to branch roots
    C. endodermis is to root hairs
    D. mesodermis is to secondary roots

    Ans. B

27. The vascular cambium of a dicot or gymnosperm stem produces two types of tissues which are referred to as:

    A. primary xylem or wood and primary phloem or inner bark
    B. secondary xylem or inner bark and secondary phloem or wood
    C. secondary xylem or inner bark and secondary periderm or outer bark
    D. secondary xylem or wood and secondary phloem

    Ans. D

28. The second lateral meristem of a dicot stem is to the collective tissues produced by the second lateral meristem, as:

    A. cork cambium is to periderm
    B. secondary vascular cambium is to cork
    C. periderm cambium is to outer bark
    D. periderm cambium is to secondary xylem and phloem

    Ans. A

29. In a dicot stem, the thin-walled, large-diameter vessels and tracheids and few fibers are to thick-walled, narrow vessels and tracheids and many fibers, as:

    A. late summer wood is to spring wood
    B. spring wood is to early summer wood
    C. spring wood is to late summer wood
    D. early summer wood is to fall-winter wood

    Ans. C

30. The movement of sugar, water, and minerals within a vascular plant is to the mechanism that is responsible for producing an upward pulling action of water, as:

    A. transpiration is to tension-cohesion mechanism
    B. translocation is to transpiration
    C. translocation is to guttation
    D. transpiration is to root pressure

    Ans. B

31. Water molecules attracted to water molecules are to water molecules attracted to the walls of the xylem cells, as:

    A. cohesion is to polar bonding
    B. hydrogen bonding is to polar bonding
    C. hydrogen bonding is to cohesion
    D. cohesion is to adhesion

    Ans. D

32. The mechanism of water transport, known as root pressure, is due to a process referred to as:

    A. transpiration
    B. translocation
    C. osmosis
    D. tension-cohesion mechanism

    Ans. C

33. The sugar transported through a plant is to the location where the sugar is stored or metabolized, as:

    A. sucrose is to sink
    B. glucose is to roots
    C. maltose is to fruit
    D. glucose is to depository

    Ans. A

34. The pressure flow hypothesis explains the movement of sugar through the plant. The correct sequence of sugar movement is:

    A. mesophyll of leaves, sieve tubes, companion cells, and depository
    B. mesophyll of leaves, companion cells, sieve tubes, and sink
    C. bundle sheath, phloem, xylem, and depository
    D. palisade, spongy, phloem, and roots

    Ans. B

Stems and Roots

35. Mineral ions travel through the root tissues are to the plant's strategy for concentrating mineral ions, as:

    A. ions pass along the cell walls with water are to selectively reject certain ions
    B. ions pass along the cell walls with water are to selectively accumulate the mineral ions they require
    C. ions travel through the root tissues from cell to cell are to selectively accumulate the mineral ions they require
    D. ions pass along the cell walls dissolved in water are to mineral ions accumulated in the same concentration they are found in the soil

    Ans. C

36. One of the most useful methods to study plant nutrition is to grow the plants in aerated water with dissolved mineral salts. This technique is referred to as:

    A. quantitative hydrodetermination
    B. hydroculture
    C. aquaculture
    D. hydroponics

    Ans. D

37. What are the advantages and disadvantages of organic and inorganic fertilizers?

38. Differentiate between the root's uptake of water from the soil and its uptake of minerals.

39. How do biologists determine which elements are essential for plant growth?

40. Trace the pathway of water from the soil through the various root tissues, stem tissues, leaf tissues, and through the leaf stomatal openings.

Chapter 29

Reproduction in Flowering Plants

1. What is the name of a seed collection, which helps preserve the genetic variation within different varieties of crops?

   A. seed bank
   B. seed herbarium
   C. seed arboretum
   D. botanical depository

   Ans. A

2. The fusion of haploid gametes is to the process in which independent assortment of genes occurs, as

   A. meiosis is to sexual reproduction
   B. sexual reproduction is to mitosis
   C. fertilization is to meiosis
   D. fertilization is to mitosis

   Ans. C

3. Process which produces haplold gametes is to one of the advantages of sexual reproduction, as

   A. mitosis is to new gene combinations
   B. syngamy is to genetic variation
   C. mitosis is to better adaptations and higher fitness
   D. meiosis is to new combinations of genes

   Ans. D

4. Process which does not usually involve the formation of flowers, seeds, and fruits is to rhizomes, tubers, bulbs, corms, and stolons, as

   A. asexual reproduction is to roots
   B. asexual reproduction is to stems
   C. syngamy is to roots
   D. photosynthesis is to roots

   Ans. B

Reproduction in Flowering Plants

5. Part of an existing plant may become separated from the rest of the plant, and subsequently grows to form a complete, independent plant is to plants produce embryos in seeds without meiosis and the fusion of reproductive cells, as

   A. grafting is to asexual reproduction
   B. asexual reproduction is to sexual reproduction
   C. asexual reproduction is to apomixis
   D. propagation is to parthenogenesis

   Ans. C

6. An example of a tuber is:

   A. crocus, and gladiolus
   B. pear, and blackberry
   C. tulips, and onions
   D. white potatoes and elephant's ear

   Ans. D

7. An example of a bulb is:

   A. lilies, and onions
   B. crocus and gladiolus
   C. strawberry, and red raspberry
   D. white potatoes and elephant's ear

   Ans. A

8. Sepals and petals are to stamens and carpels, as

   A. fertile modified leaves are to sterile modified leaves
   B. sterile modified leaves are to fertile modified stems
   C. sterile modified leaves are to fertile modified leaves
   D. sterile modified epidermal cells are to fertile modified parenchyma cells

   Ans. C

9. "Male" organs are to "female" organs, as:

   A. carpels are to stamens
   B. stamens are to carpels
   C. carpels are to sepals
   D. sepals are to ovules

   Ans. B

10. A flower that possesses both stamens and carpels is to a flower that does not have both stamens and carpels, as:

    A. perfect is to imperfect
    B. complete is to perfect
    C. complete is to incomplete
    D. heterogenous is to homogenous

    Ans. A

11. A collective term for all sepals is to a collective term for all the petals, as:

    A. corolla is to pistil
    B. sepolla is to petolla
    C. pistil is to calyx
    D. calyx is to corolla

    Ans. D

12. Components or anatomical parts of stamens are to anatomical parts or components of a carpel, as:

    A. stigma and antlers are to filaments and ovules
    B. stigma and ovules are to styles and anthers
    C. filaments and anthers are to stigma, style, and ovary
    D. ovary and anthers are to stigma, style, and filaments.

    Ans. C

13. Each pollen grain produces how many sperm?

    A. one
    B. two
    C. four
    D. eight

    Ans. B

14. How many haploid megaspore are initially produced?

    A. one
    B. two
    C. four
    D. eight

    Ans. C

Reproduction in Flowering Plants

15. The female embryo sac consists of how many haploid nuclei?

    A. one
    B. two
    C. four
    D. eight

    Ans. D

16. Male gametophyte is to female gametophyte, as:

    A. pollen grain is to embryo sac
    B. sperm is to ovule
    C. semen is to ovary
    D. sperm is to megaspore

    Ans. A

17. The transfer of pollen to the "female" portion of the flower, is referred to as:

    A. fertilization
    B. pollination
    C. syngamy
    D. sexual reproduction

    Ans. B

18. Animal pollinators have been a strong selective force in the evolution of certain features of flowering plants, and plants have likewise affected the evolution of certain features of their animal pollinators. When two different organisms interact so closely that they become increasingly adapted to one another, the process is referred to as

    A. selective evolution
    B. natural selection
    C. coevolution
    D. fitness

    Ans. C

19. Flowers that are red, orange, or yellow and lack a scent are to dull white flowers that possess a strong scent, as:

    A. insect pollinated flowers are to bird pollinated flowers
    B. bird pollinated flowers are to bat pollinated flowers
    C. wind pollinated flowers are to insect pollinated flowers
    D. wind pollinated flowers are to bat pollinated flowers

    Ans. B

20. One of the rewards for the animal pollinator is food. A sugary solution used as an energy-rich food is to a protein rich food, as:

    A. nectar is to ovules
    s. nectar is to magaspore
    C. embryo sac is to microspores
    D. nectar is to pollen

    Ans. D

21. Flowers that are often small and inconspicuous are pollinated by

    A. wind
    B. small gnatlike insects
    C. nocturnal or night insects
    D. moths and butterflies

    Ans. A

22. The egg within the embryo sac fuses with one sperm, which will develop into what structure within the seed?

    A. endosperm
    B. endometrium
    C. embryonic plant
    D. megaspore mother cell

    Ans. C

23. The two polar nuclei fuse with a sperm, to form what structure within the seed?

    A. megaspore mother cell
    B. endosperm
    C. embryo sac
    D. embryonic plant

    Ans. B

24. The process, in which two separate cell fusions occur in flowering plants, is referred to as:

    A. diploid sexual reproduction
    B. dual fertilization
    C. double diploid fertilization
    D. double fertilization

    Ans. D

Reproduction in Flowering Plants

25. The two cells formed as a result of the first division of the fertilized egg establish polarity, or direction, in the embryo. The bottom cell typically develops into a structure referred to as:

    A. proembryo
    B. postembryo
    C. suspensor
    D. seed embryo

    Ans. C

26. The two cells formed as a result of the first division of the fertilized egg establish polarity, or direction, in the embryo. The bottom cell functions in:

    A. anchoring the embryo and aids in nutrient uptake from the endosperm
    B. developing into the actual embryo
    C. developing into the cotyledon and mitotically developing the first leaves
    D. developing into the cotyledons and absorbing stored nutrients from the developing root

    Ans. A

27. The two cells formed as a result of the first division of the fertilized egg establish polarity, or direction, in the embryo. The top cell divides to form a short chain of cells, referred to as

    A. globular embryo
    B. torpedo embryo
    C. postembryo
    D. proembryo

    Ans. D

28. Food storage organs found within seeds are to a tough, protective outer layer of a seed, as:

    A. suspensor and globular embryo are to seed wall
    B. cotyledon and endosperm are to seed coat
    C. starch and lipids (oils) are to seed integument
    D. hypocotyl and radical are to seed integument

    Ans. B

29. Embryonic root, embryonic shoot, and one or two cotyledons are to the structure following fertilization that the ovary develops into, as:

    A. endosperm is to mature embryo
    B. mature embryo is to endosperm
    C. mature embryo is to fruit
    D. mature embryo is to seed coat

    Ans. C

30. A simple fruit such as a tomato or banana is to a simple fruit such as a peach or avocado, as:

    A. berry is to drupe
    B. samara is to pome
    C. pepo is to capsule
    D. hesperidium is to achene

    Ans. A

31. Fruits such as raspberries and blackberries are to fruits such as pineapple, as:

    A. berry is to pome
    B. aggregate fruit is to multiple fruit
    C. berry is to drupe
    D. berry is to pepo

    Ans. B

32. Milkweed pods and pea pods at maturity are to strawberries and apples, as:

    A. grains are to aggregate fruits
    B. multiple fruits are to berries
    C. dry fruits are to multiple fruits
    D. dry fruits are to accessory fruits

    Ans. D

33. What type of seed disposal mechanisms do maples, dandelions, and milkweeds exhibit?

    A. animal
    B. explosive dehiscence
    C. wind
    D. water

    Ans. C

Reproduction in Flowering Plants

34. What type of adaptive dispersal mechanism do seeds that have spines and barbs exhibit?

A. animal
B. explosive dehiscence
C. wind
D. water

Ans. A

35. The seed disposal mechanism of the coconut is to the seed dispersal mechanism of the touch-me-not and bitter cress, as:

A. animal is to wind
B. water is to explosive dehiscence
C. water is to animal
D. wind is to water

Ans. B

36. The site where pollen is initially transferred or received on the flower is to the structure in which the pollen tube grows, as:

A. style is to ovary
B. embryo sac is to ovary
C. stigma is to style
D. receptacle is to embryonic sac

Ans. C

37. Differentiate between sexual and asexual reproduction.

38. Differentiate between pollination and fertilization. Explain each process.

39. Differentiate between seeds, and fruits. Explain the process by which each is produced.

40. Explain the roles of each of the following in the life cycle of a flowering plant.
a. meiosis
b. mitosis
c. pollen tube
d. embryo sac
e. ovary
f. double fertilization
g. endosperm

Chapter 30

Regulation of Plant Growth and Development

1. Agents white, blue, orange are classified as:

   A. herbicides
   B. pesticides
   C. fungicides
   D. bacteriacides

   Ans. A

2. Dioxin, 2, 4-D and 2,4,5-T are compounds specifically associated with:

   A. auxins
   B. DDT
   C. Agent Orange
   D. funginex

   Ans. C

3. Typically, the region in which the initiation of sexual reproduction is often under environmental control is to the factors that influence flowering as:

   A. tropics are to daylength and temperature
   B. temperate latitudes are to daylength and temperatures
   C. arctic latitudes are to temperature and water availability
   D. deserts are to temperature and daylength

   Ans. B

4. The response of a plant to the relative lengths of daylight and darkness, is referred to as:

   A. temporal response
   B. diurnal-nocturnal response
   C. temporal regulatory mechanisms
   D. photoperiodism

   Ans. D

5. When the night length is equal to or greater than some critical length is to when the night length is equal to or less than some critical length, as:

   A. long-day plants are to short-day plants
   B. short-day plants are to long-day plants
   C. short-day plants are to day-neutral plants
   D. day-neutral plants are to long-day plants

   Ans. B

6. Plants that do not initiate flowering in response to seasonal changes in daylength, are referred to as:

   A. long-day plants
   B. short-day plants
   C. day-neutral plants
   D. phototrophic plants

   Ans. C

7. Plants that are able to detect the shortening days and lengthening nights are to plants that are able to detect the lengthening days and shortening nights, as:

   A. long-day plants are to short-day plants
   B. short-day plants are to day-neutral plants
   C. long-day plants are to day-neutral plants
   D. short-day plants are to long-day plants

   Ans. D

8. Plants that typically flower late in summer or fall, are referred to as:

   A. short-day plants
   B. day-neutral plants
   C. estival plants
   D. long-day plants

   Ans. A

9. Plants that typically flower in the spring and early summer, are referred to as:

   A. short-day plants
   B. estival plants
   C. prevernal plants
   D. long-day plants

   Ans. D

10. For plants, to have a biological response to light there must be a mechanism to absorb light, that is referred to as:

    A. chlorophyll
    B. stigma
    C. photoreceptor
    D. photon initiator

    Ans. C

11. The pigment responsible for photoperiodism and a number of other light-initiated responses of plants is a blue-green pigment, referred to as:

    A. chlorophyll
    B. phytochrome
    C. xanthochrome
    D. photochrome

    Ans. B

12. A plant pigment that increases when the plant is exposed to sunlight, and slowly decreases at night, is referred to as:

    A. chlorophyll
    B. Pr
    C. Pfr
    D. Pao

    Ans. C

13. What substance inhibits flowering in short-day plants?

    A. Pr
    B. Pao
    C. Pir
    D. Pfr

    Ans. D

14. Short-day plants will not flower if they are interrupted at night by what color of light?

    A. red light
    B. blue light
    C. violet light
    D. green light

    Ans. A

Regulation of Plant Growth and Development

15. If short-day plants are interrupted by a burst of a certain color of light at night, certain plant pigments are converted from one form to another. Which pigments are involved in this conversion?

    A. Pao to Pfr
    B. Pfr to Pr
    C. Pir to Pr
    D. Pr to Pfr

    Ans. D

16. What substance induces flowering in long-day plants?

    A. Pr
    B. Pir
    C. Pif
    D. Pfr

    Ans. D

17. The promotion of flowering by exposure to low temperature for a period of time is known as:

    A. estivalization
    B. vernalizatlon
    C. thermalizatlon
    D. temporal regulation

    Ans. B

18. An absolute requirement for germination is to the source of energy for germination as:

    A. warm temperatures are to starch
    B. increasing daylength is to glucose
    C. water is to ATP
    D. increasing daylength is to starch

    Ans. C

19. What factor is typically required for the germination of tiny seeds?

    A. low levels of precipitation
    B. red light
    C. blue light
    D. violet light

    Ans. B

20. What factor ensures that seeds germinate in the spring rather than in the winter?

    A. prolonged cold periods
    B. large amounts of precipitation
    C prolonged warm periods
    D. alternating warm and cold periods

    Ans. A

21. Natural chemical inhibitor of germination is to a factor that alleviates the chemical inhibitor, as:

    A. auxin is to sunlight
    B. auxin is to warm temperature
    C. abscisic acid is to rain
    D. indoleacetic acid is to sunlight

    Ans. C

22. The first part of the plant to emerge from the seed during germination is:

    A. embryonic stem
    B. coleoptile
    C. cotyledon
    D. embryonic root

    Ans. D

23. Monocots have a protective sheath of cells that protect the young shoot as it pushes upward through the soil, which is referred to as:

    A. coleoptile
    B. radical
    C. cotyledon
    D. epicotyl

    Ans. A

24. Plants, animals, and micro-organisms appear to have an internal time, or biological clock, that approximates a 24-hour cycle. These internal cycles, are referred to as:

    A. bio-feedback mechanisms
    B. circadian rhythms
    C. diurm rhythms
    D. bio-rhythms

    Ans. B

Regulation of Plant Growth and Development

25. Two examples of biological clock activity in plants are:

    A. opening and closing of stomata and water absorption
    B. flowering and germination
    C. flowering and phototropism
    D. sleep movements and opening and closing of stomata

    Ans. D

26. In plants changes in turgor can induce:

    A. flowering
    B. seed development
    C. temporary movement
    D. leaf development

    Ans. C

27. Solar tracking is frequently observed in:

    A. leaves or flowers
    B. seeds or leaves
    C. flowers or stems
    D. stems or leaves

    Ans. A

28. The growth of a plant in response to light is to the growth of a plant in response to a mechanical stimulus, as:

    A. phototropism is to gravitropism
    B. phototropism is to galvanotropism
    C. phototropism is to chemotropism
    D. phototropism is to thigmotropism

    Ans. D

29. What is a chemical messenger produced in one part of a plant and transported to another part, where it elicits a physiological response called?

    A. tropism
    B. hormone
    C. biological stimulatory substance
    D. chemical activator

    Ans. B

30. A plant substance that promotes cell growth by triggering cell elongation is to a plant substance that promotes cell division and differentiation, as:

    A. auxins are to indoleacetic acid
    B. cytokinins are to ethylene
    C. auxins are to cytokinins
    D. abscisic acids are to indoleacetic acid

    Ans. C

31. A plant substance that promotes apical dominance is to a plant substance that promotes growth, as:

    A. auxins is to gibberellins
    B. abscisic acid is to ethylene
    C. gibberellins are to abscisic acid
    D. ethylene is to indoleacetic acid

    Ans. A

32. What substance will result in the development of seedless fruit, when it is applied to flowers in which fertilization has not occurred?

    A. ethylene
    B. gibberellin
    C. abscisic acid
    D. auxin

    Ans. D

33. What ingredient is required for plant tissue culture to be successful?

    A. auxin
    B. cytokinins
    C. ethylene
    D. gibberellins

    Ans. B

34. The relationship is antagonistic between these two plant substances, concerning the growth of lateral buds:

    A. auxin and gibberellins
    B. cytokinins and abscisic acid
    C. auxin and cytokinins
    D. auxin and ethylene

    Ans. C

Regulation of Plant Growth and Development

35. Ripening of fruit is to promotion of seed dormancy as:

    A. ethylene is to abscisic acid
    B. auxin is to ethylene
    C. auxin is to gibberellins
    D. cytokinins are to auxins

    Ans. A

36. A rotten apple or pear is to the formation of bud scales covering the terminal buds, as;

    A. auxins is to abscisic acid
    B. gibberellins are to ethylene
    C. cytokinins are to auxin
    D. ethylene is to abscisic acid

    Ans. D

37. Explain why plant growth and development is so sensitive to environmental factors or cues.

38. Explain the role of phytochrome in the flowering process.

39. Explain what factors influence the germination of seeds, and why the plant responds the way it does to those factors.

40. Explain how auxin is involved in phototropism.

Chapter 31

Animal Tissues, Organs, and Organ Systems

1. Arrange the following in order of increasing complexity:
   1. Organ
   2. Cell
   3. System
   4. Tissue

   A. 2,4,1,3
   B. 2,4,3,1
   C. 3,1,4,2
   D. 1,2,3,4

   Ans. A

2. A group of closely associated cells that are adapted to carry out a specific function is called a:

   A. tissue
   B. organ
   C. system
   D. organism

   Ans. A

3. The epithelial layer of the skin is called the:

   A. basement membrane
   B. dermis
   C. epidermis
   D. psuedostratified layer

   Ans. C

4. The basement membrane of epithelial tissues is:

   A. a layer of very flat cells
   B. a membranous film under the epithelial cells
   C. tiny fibers embedded in polysaccharides
   D. tiny fibers embedded on an intracellular protein

   Ans. C

5. There are three types of epithelial cells. The flattened type of epithelial cell is called a

   A. squamous cell
   B. cuboidal cell
   C. columnar cell
   D. goblet cell

   Ans. A

6. The elongated column-like epithelial cell is a

   A. cuboidal cell
   B. columnar cell
   C. squamous cell
   D. goblet cell

   Ans. B

7. Some epithelial linings of cavities and passageways produce mucus. This mucus is produced by:

   A. cuboidal cells
   B. goblet cells
   C. columnar cells
   D. squamous cells

   Ans. B

8. Glands contain epithelial cells. Glands which secrete their products on to a free epithelial surface, through a duct, are:

   A. exocrine glands
   B. endocrine glands
   C. thyroid glands
   D. mucous glands

   Ans. A

9. Fibers found in connective tissue are produced by:

   A. goblet cells
   B. fibroblasts
   C. adipose cells
   D. epithelial cells

   Ans. B

10. The fiber which snaps back to the normal length when stress is removed, is called a(n):

    A. collagen fiber
    B. elastic fiber
    C. reticular fiber
    D. muscle fiber

    Ans. B

11. Fine fibers, which branch and form the support network within many tissues and organs, are called:

    A. elastic fibers
    B. collagen fibers
    C. muscle fibers
    D. reticular fibers

    Ans. D

12. Collagen bundles arranged in a definite pattern making the tissue resistant to stress are:

    A. loose connective tissue
    B. dense connective tissue
    C. adipose tissue
    D. muscle tissue

    Ans. B

13. Adipose tissue stores:

    A. protein
    B. carbohydrates
    C. fats
    D. water

    Ans. C

14. How does adipose tissue differ from other connective tissues?

    A. cellular products of the connective tissue are deposited outside the cell
    B. cellular products of the connective tissue cell are kept inside the cell
    C. the matrix is produced somewhere else and deposited in the connective tissue
    D. the matrix of all connective tissue has the same origin

    Ans. B

15. Chondrocytes found in cartilage are housed in small cavities called:

    A. lumen
    B. lacunae
    C. osteon
    D. lamellae

    Ans. B

16. Which of the following best describes the functional organization of cartilage?

    A. it has many nerves in the tissue
    B. it has blood vessels distributed extensively in the tissue
    C. nutrient is distributed to the cells only by diffusion
    D. lymph vessels effectively drain tissue fluid from the cartilage

    Ans. C

17. The concentric layer (rings) of bone matrix in compact bone are called:

    A. lacunae
    B. lamellae
    C. Haversian canal
    D. osteon

    Ans. B

18. The structural unit of compact bone is the __________, which contain the cells of bone called __________.

    A. lacunae, adipocytes
    B. lamellae, chondrocytes
    C. Haversian canal, osteoblasts
    D. osteon, osteocyte

    Ans. D

19. Blood and lymph are best described as:

    A. cellular tissue
    B. circulating tissue
    C. dense tissue
    D. fixed fibrous tissue

    Ans. B

20. Which of the following tissues are not specialized for contraction?

    A. smooth muscle
    B. cardiac muscle
    C. skeletal muscle
    D. elastic cartilage

    Ans. D

21. Which of the following tissues has the most extensive striations?

    A. cardiac muscle
    B. skeletal muscle
    C. smooth muscle
    D. visceral muscle

    Ans. B

22. Glial cells are found in:

    A. muscle tissue
    B. blood tissue
    C. nervous tissue
    D. connective tissue

    Ans. C

23. An example of positive feedback is:

    A. regulating blood sugar
    B. delivery of a baby
    C. regulation of body temperature
    D. producing urine

    Ans. B

24. Lining of the digestive tract is:

    A. simple squamous epithelium
    B. simple cuboidal epithelium
    C. simple columnar epithelium
    D. psuedostratified epithelium

    Ans. C

25. Respiratory passages are lined with:

    A. stratified squamous epithelium
    B. psuedostratified columnar epithelium
    C. stratified cuboidal epithelium
    D. simple squamous epithelium

    Ans. B

26. Tissue that forms pads around certain internal organs is:

    A. adipose tissue
    B. loose connective tissue
    C. cartilage tissue
    D. bone

    Ans. A

27. Which statement best describes skeletal muscle?

    A. multinucleated with nuclei peripherally located
    B. multinucleated with nuclei centrally located
    C. single nucleus centrally located
    D. two nuclei located peripherally

    Ans. A

28. Which of the following is not a homeostatic function of the respiratory system?

    A. maintain oxygen level
    B. regulate blood pH
    C. ionic balance
    D. eliminates carbon dioxide

    Ans. C

29. What tissue looks like the cross section of several tree trunks?

    A. adipose tissue
    B. compact bone
    C. skeletal muscle
    D. cartilage muscle

    Ans. B

30. Type of muscle which lacks striation

A. smooth muscles
B. skeletal muscles
C. cardiac muscles
D. voluntary muscle

Ans. A

31. What are homeostatic mechanisms called?

A. negative feedback system
B. Biofeedback system
C. positive feedback system
D. equilibrium

Ans. B

32. Intercalated discs are found in what type of tissue?

A. bone tissue
B. skeletal muscle
C. cardiac muscle
D. adipose tissue

Ans. C

33. What system besides the endocrine system, helps maintain homeostatic levels of calcium?

A. nervous system
B. skeletal system
C. urinary system
D. muscular system

Ans. B

34. Which tissue has a fluid intercellular matrix?

A. adipose tissue
B. blood tissue
C. cartilage tissue
D. epithelial tissue

Ans. B

Animal Tissues, Organs, and Organ Systems

35. Haversian canals serve as passages for blood vessels to transport nutrients to:

    A. compact bone
    B. lacunae
    C. osteocytes
    D. lamellae

    Ans. C

36. Insulation is a function of:

    A. blood tissue
    B. skin
    C. cartilage
    D. adipose tissue

    Ans. D

37. Name the various tissues you would expect to find in the stomach and give the function of each tissue.

38. Which of the body systems has the least to do with homeostasis and explain how this fits into the scheme of organizational function.

39. In all reality, what is the nature of homeostasis and equilibrium?

40. Why are negative feedback mechanisms more common than positive feedback mechanisms and what terminates a positive feedback mechanism?

Chapter 32

Skin, Bones, and Muscle: Protection, Support, and Locomotion

1. Which of the following best contrasts stratum basale and the stratum corneum?

   A. live cells, dead cells
   B. outer layer, inner layer
   C. dermis, epidermis
   D. mitosis, meiosis

   Ans. A

2. Which of the following layers of the skin is fibrous connective tissue composed mainly of collagen fibers?

   A. epidermis
   B. dermis
   C. stratum basale
   D. stratum corneum

   Ans. B

3. The molecule that gives the skin its mechanical strength, flexibility, and waterproof quality is

   A. sebum
   B. keratin
   C. mucous
   D. pangolin

   Ans. B

4. The layer of the skin which is composed of layers of keratin producing squamous cells is the:

   A. epidermis
   B. dermis
   C. stratum basale
   D. cuticle

   Ans. A

5. The dermis contains all but which of the following structures?

   A. fatty tissue
   B. sweat glands
   C. hair follicles
   D. sense organs

   Ans. A

6. Muscles deliver their force through another medium to do work. In some of the simplest of animals, muscles act through:

   A. endoskeleton
   B. exoskeleton
   C. hydroskeleton
   D. cuticle

   Ans. C

7. A lifeless deposit of calcium-impregnated tissue atop the epidermis best describes a(n):

   A. cuticle
   B. endoskeleton
   C. exoskeleton
   D. hydroskeleton

   Ans. C

8. Which of the following best contrasts an exoskeleton and an endoskeleton?

   A. nonliving, living
   B. calcium, iron
   C. fixed, jointed
   D. regenerative, permanent

   Ans. A

9. Which best describes movement in a hydrostatic skeleton?

   A. crude mass movement of body
   B. protect body from outside forces
   C. fine movement of appendages
   D. forces transmitted generally in one direction

   Ans. A

10. All animals demonstrate some form of hydrostatic support. How is hydrostatic support illustrated in mammals?

    A. cranium around the brain
    B. swelling at an injured point
    C. erection of sensitive tissues in sexual arousal
    D. pressure in the blood vessels

    Ans. C

11. Which best illustrates the intensity of hydrostatic forces in animals?

    A. extension of a hydra body
    B. extension of earth worm through soil
    C. clam extends foot
    D. sea stars extending their tube feet

    Ans. B

12. Which of the following best contrasts the exoskeleton of a mollusc and arthropod?

    A. jointed; unjointed
    B. permanent; temporary
    C. calcium; no calcium, but protein
    D. support; protection

    Ans. B

13. Which skeleton provides for the greatest variety of motions?

    A. hydrostatic skeleton
    B. external skeleton
    C. internal skeleton
    D. cuticle

    Ans. C

14. With respect to the skeleton, what makes sharks different from other animals?

    A. they have no skeleton
    B. they have a cartilgenous skeleton
    C. they have an exoskeleton
    D. since sharks are found in salt water, the buoyancy of the water supports their body

    Ans. B

15. Which section of the vertebral column contains the greater number of vertebrae?

    A. cervical
    B. thoracic
    C. lumbar
    D. sacral

    Ans. B

16. How many ribs are attached to the sternum?

    A. 7
    B. 5
    C. 3
    D. 2

    Ans. A

17. When comparing the pectoral skeleton of a variety of vertebrates, the most variable parts are:

    A. shoulder bones
    B. arm bones
    C. forearm bones
    D. finger bones

    Ans. D

18. What part of the bone contributes to the lever action of a bone and amplifies the motion generated by the muscles?

    A. epiphysis
    B. diaphysis
    C. periosteum
    D. epiphyseal plate

    Ans. B

19. Name the growth plate of a long bone.

    A. epiphysis
    B. diaphysis
    C. epiphyseal plate
    D. periosteum

    Ans. C

20. The spindle-shaped units of dense bone are called:

    A. lacunae
    B. osteons
    C. Haversian canals
    D. spongy bone

    Ans. B

21. Cells that break down bone are:

    A. osteocytes
    B. osteoblast
    C. osteoclast
    D. lacunae

    Ans. C

22. Bone cells trapped in the bone matrix are called:

    A. osteocytes
    B. osteoblasts
    C. osteoclasts
    D. lacunae

    Ans. A

23. What component of bone contributes to bone flexibility?

    A. mineral of bone
    B. collagen fibers
    C. hard matrix
    D. lacunae

    Ans. B

24. The joint capsule which is lined with a membrane that secretes a lubricant called synovial fluid is associated with:

    A. immovable joints
    B. slightly movable joints
    C. freely movable joints
    D. common joint disorders

    Ans. C

25. The thick myofilaments are:

    A. myosin
    B. actin
    C. myofibril
    D. sarcomere

    Ans. A

26. The 2 lines bound the:

    A. sarcoplasm
    B. sarcoplasmic reticulum
    C. striations
    D. sarcomere

    Ans. D

27. The force to slide the actin myofilaments over the myosin is generated by:

    A. myosin splitting ATP
    B. calcium released in sarcoplasm
    C. flexion of the crossbridges
    D. uncovers binding sites of the actin filaments

    Ans. C

28. Which of the following occurs first?

    A. action potential spreads through tubes
    B. acetylcholine binds the receptor
    C. motor neuron releases acetylcholine
    D. cross bridges flex

    Ans. C

29. What causes rigor mortis?

    A. lack of oxygen
    B. ATP depletion
    C. fatigue
    D. excess of CO_2

    Ans. B

30. Energy stored in muscles is stored in the form of:

A. ATP
B. creatin phosphate
C. ADP
D. glycogen

Ans. B

31. Muscle fatigue:

A. is the result of buildup of lactic acid
B. occurs when muscle gets tired
C. is aerobic metabolism
D. occurs when oxygen is not available

Ans. A

32. What is the relationship of the triceps to the biceps?

A. agonist
B. antagonist
C. complement
D. articulates

Ans. B

33. A simple quick contraction of a muscle fiber is called a:

A. muscle contraction
B. simple muscle twitch
C. tetanus
D. summation

Ans. B

34. Smooth muscle contraction

A. single, sustained, smooth contraction
B. contracts abruptly and rhythmically
C. contracts slowly
D. is a tetanic contraction

Ans. C

35. Muscles protect themselves from oxygen shortage by producing:

    A. glycogen
    B. myoglobin
    C. fast twitch fibers
    D. slow twitch fibers

    Ans. B

36. Which is larger?

    A. myofibril
    B. myofilament
    C. muscle fiber
    D. myosin

    Ans. C

37. Since a muscle shortens by the shortening of the sarcomere when myofilaments slide past each other, how does a muscle lengthen to contract again?

38. Knowing how the cross bridges attach and disconnect from the actin, why is rigor mortis temporary and why do the dead become very flexible?

39. Explain how an earthworm moves forward.

40. What are the advantages of transdermal patch technology?

Chapter 33

Responsiveness: Neural Control

1. Arrange the elements of a neural response, listed below, in appropriate order from the point the stimulus is delivered:
   1. integration
   2. reception
   3. transmission
   4. response

   A. 4,3,1,2
   B. 2,3,1,4
   C. 2,3,4,1
   D. 3,2,1,4

   Ans. B

2. The response of the nervous system is expressed as:

   A. transmission of impulse to CNS
   B. release of a neurotransmitter
   C. the contraction of a muscle or action of a gland
   D. activation of a receptor

   Ans. C

3. Cells that remove debris from nervous tissue are:

   A. neuron
   B. phagocytic glial cells
   C. neuroglia cells
   D. Schwann cells

   Ans. B

4. A neuron consists of all the following except:

   A. synapse
   B. axon
   C. dendrite
   D. cell body

   Ans. A

Responsiveness: Neural Control

5. The gaps between Schwann cells are called:

   A. synapse
   B. nodes of Ranvier
   C. myelin sheath
   D. synaptic knobs

   Ans. B

6. The myelin sheath is formed by:

   A. neurons
   B. axon of a neuron
   C. Schwann cells
   D. neuroglia

   Ans. C

7. Nerve cell bodies clustered together in a mass are called a:

   A. nerve
   B. myelin sheath
   C. ganglion
   D. nucleus

   Ans. C

8. The cytoplasm of a neuron is __________, relative to the interstitial fluid.

   A. positively charged
   B. negatively charged
   C. neutral
   D. resting

   Ans. B

9. The resting membrane potential is:

   A. +70 millivolts
   B. -70 millivolts
   C. +30 millivolts
   D. 0 millivolts

   Ans. B

10. The sodium-potassium pump keeps:

    A. K+ concentration inside 30 times the outside
    B. Na+ concentration inside 14 times greater than outside
    C. K+ concentration outside is 30 times greater than the inside
    D. K+ concentration inside and Na+ concentration outside are equal

    Ans. A

11. The negative charge inside the neuron is the result of:

    A. lack of Na+ ions
    B. negative charge proteins and organic phosphates
    C. low concentration of K+ ions
    D. high concentration of K+ ions outside the cell

    Ans. B

12. The critical voltage point that opens the voltage-activated ion channels is called:

    A. wave of depolarization
    B. threshold level
    C. repolarization
    D. refractory period

13. Which of the following are in proper sequence?

    A. repolarization, depolarization
    B. K+ voltage channels open, Na+ voltage channels open
    C. Na+ voltage channels open, K+ voltage channels open
    D. refractory, repolarize

    Ans. C

14. Arrange these events in order as they would occur in an action potential.
    1. membrane depolarizes
    2. K+ voltage sensitive channels open
    3. Na+ voltage sensitive channels open
    4. membrane repolarizes

    A. 3,1,2,4
    B. 2,3,1,4
    C. 1,2,3,4
    D. 2,1,3,4

    Ans. A

15. Which of the following best contrasts saltatory and continuous conduction?

    A. unmyelinated, myelinated
    B. slow, fast
    C. more efficient, less efficient
    D. more ions pumped by sodium-potassium pump, fewer ions need to be pumped by Na-K pump

    Ans. C

16. The all-or-none-law states the neuron:

    A. depolarizes completely at the threshold of the nerve
    B. has a grade action potential
    C. maintains the resting membrane potential
    D. will depolarize with stimulus less than threshold

    Ans. A

17. Gap junctions between neurons form what type of synapse?

    A. chemical synapse
    B. synaptic cleft
    C. electrical synapse
    D. synaptic bulb

    Ans. C

18. A chemical messenger which diffuses across a synapse to activate the postsynaptic neuron is called:

    A. hormone
    B. neurohormone
    C. neurotransmitter
    D. prostaglandin

    Ans. C

19. When a postsynaptic neuron is partially depolarized, the synapse is referred to as:

    A. inhibitory synapse
    B. excitatory synapse
    C. electrical synapse
    D. chemical synapse

    Ans. B

20. When potassium ions leave the postsynaptic neuron and the interior of the neuron becomes more negative relative to the surrounding fluid, the neuronal membrane is said to be

A. hyperpolarized
B. hypopolarized
C. depolarized
D. repolarized

Ans. A

21. When the neurotransmitters hyperpolarize the postsynaptic neuronal membrane, an action potential is less likely to form. This synapse is called:

A. an electrical synapse
B. a chemical synapse
C. an excitatory synapse
D. an inhibitory synapse

Ans. D

22. Neurons which release acetylcholine as a neurotransmitter are called:

A. adrenergic neurons
B. unipolar neurons
C. motor neurons
D. cholinergic neurons

Ans. D

23. Which of the following best reflects the action of acetylcholine?

A. excitatory effect on cardiac muscle
B. excitatory effect on skeletal muscle
C. increases the heart rate
D. inhibitory effect on skeletal muscle

Ans. B

24. Which of the following reflects the fate of acetylcholine in the synapse?

A. it binds to receptors on the presynaptic neuron
B. it is washed away in tissue fluid
C. it is destroyed by acetylcholinesterase
D. it simply stays in the synapse to continue synaptic transmission

Ans. C

Responsiveness: Neural Control

25. Norepinephrine is a neurotransmitter released by:

    A. cholinergic neurons
    B. adrenergic neurons
    C. sensory neurons
    D. all motor neurons

    Ans. B

26. Mood altering drugs work by changing the levels of what chemical messengers in the brain?

    A. neurohormones
    B. acetylcholine
    C. monoamine oxidase
    D. catecholamines

    Ans. D

27. The neural pathways of impulse transmission are one way; which of the following assure the one way transmission?

    A. axon, synapse, and dendrite
    B. the neuron
    C. synapse and dendrites
    D. dendrites and axons

    Ans. A

28. Which of the following structures has the least effect on the rate of transmission?

    A. Na-K pump
    B. diameter of the axon
    C. the amount of myelin
    D. the spacing of the nodes of Raniver

    Ans. A

29. What parts of the neuron are responsible for integrating the hundreds of impulses that come to a neuron?

    A. cell body and axon
    B. dendrite and axon
    C. dendrite and cell body
    D. only the cell body

    Ans. C

30. A change in the potential that brings the neuron closer to firing is called the:

    A. threshold
    B. action potential
    C. resting membrane potential
    D. excitatory postsynaptic potential

    Ans. D

31. A change in the potential that brings the neuron farther from firing is called the:

    A. excitatory postsynaptic potential
    B. inhibitory postsynaptic potential
    C. action potential
    D. hypopolarization

    Ans. B

32. Arrange the following in proper order of a reflex action.
    1. association neuron
    2. sensory neuron
    3. receptor
    4. motor neuron
    5. effector

    A. 3,2,1,5,4
    B. 1,3,2,5,4
    C. 3,2,1,4,5
    D. 4,5,3,2,1

    Ans. C

33. A relatively fixed reaction pattern to a simple stimulus is a:

    A. muscle twitch
    B. reflex action
    C. reflex arc
    D. pain

    Ans. B

34. How many neurons are required to carry a response to a stimulus?

    A. 2
    B. 3
    C. 4
    D. 5

    Ans. B

Responsiveness: Neural Control

35. The neural association where a single presynaptic neuron stimulates many postsynaptic neurons is called:

    A. convergence
    B. divergence
    C. reverberating circuit
    D. facilitation

    Ans. B

36. Which of the following best contrasts convergence and divergence?

    A. merge and separate
    B. separate and merge
    C. ground and transmit
    D. linear and branching

    Ans. A

37. Explain the functional and adaptive significance of the fact that the nervous system is fast to respond but short-lived while the endocrine system is slow to respond and the response is long-lived.

38. Since an action potential can move in any direction on the neuron, how come the nervous system is unidirectional and what structural and functional component assures the direction the impulse will travel?

39. Explain why when you touch a hot stove, you pull your hand back almost before you realize you burned your hand. Explain the neural mechanisms of the response.

40. Explain the movement of ions across the cell membrane on an electrical gradient and how the voltage-sensitive channels opening and closing direct the flow of ions in an action potential.

Chapter 34

Responsiveness: Nervous Systems

1. The hydra has a nervous system to aid in the capture of prey which is best described as

   A. a nerve net
   B. having no nervous system
   C. bilateral nervous system
   D. a central nervous system

   Ans. A

2. Listed below are some trends in the evolution of the nervous system. Arrange them in order of occurrence.
   1. increased number of associate neurons
   2. increased number of neurons
   3. specialization of functions
   4. cephalization
   5. concentration of neurons in the brain

   A. 2,5,1,3,4
   B. 2,5,3,1,4
   C. 3,2,5,1,4
   D. 5,3,2,1,4

   Ans. B

3. Which of the following best describes the brain of an earthworm?

   A. cerebral ganglion
   B. segmentally arranged ganglion
   C. three pairs of ganglia
   D. nerve net

   Ans. B

4. Which of the following is NOT a characteristic of a vertebrate nervous system?

   A. ventral nerve cord
   B. dorsal nerve cord
   C. hollow nerve cord
   D. well-developed brain

   Ans. A

5. Divisions of the nervous system of a vertebrate are:

   A. autonomic and somatic
   B. central and peripheral
   C. sensory and motor
   D. parasympathetic and sympathetic

   Ans. B

6. Subdivisions of the peripheral nervous system are:

   A. CNS and PNS
   B. sympathetic and parasympathetic
   C. somatic and visceral
   D. sensory and motor

   Ans. C

7. The autonomic system has two divisions. They are:

   A. somatic and visceral
   B. sensory and motor
   C. CNS and PNS
   D. sympathetic and parasympathetic

   Ans. D

8. Afferent nerves are:

   A. symphthetic
   B. somatic
   C. sensory
   D. motor

   Ans. C

9. Muscle coordination is centered in what part of the brain?

   A. medulla
   B. cerebellum
   C. hindbrain
   D. pons

   Ans. B

10. Swallowing, vomiting, and coughing are regulated by the:

A. medulla
B. cerebrum
C. cerebellum
D. pons

Ans. A

11. The red nucleus is a nucleus which integrates information about muscle tone and posture and is located in the:

A. cerebrum
B. cerebellum
C. midbrain
D. medulla

Ans. C

12. The portion of the forebrain which relays motor and sensory messages to the cerebrum is:

A. the hypothalamus
B. thalamus
C. pituitary
D. cerebrum

Ans. B

13. The brain stem includes all of the following except:

A. medulla
B. pons
C. thalamus
D. cerebrum

Ans. D

14. The hypothalamus controls what organ with hormones?

A. anterior pituitary
B. pons
C. thyroid gland
D. cerebellum

Ans. A

Responsiveness: Nervous Systems

15. The gray matter of the cerebrum is called:

    A. brain stem
    B. cerebral hemisphere
    C. cerebral cortex
    D. pons

    Ans. C

16. In small mammals the brain is smooth while large animals have convolutions on the surface with sulci alternated with:

    A. white mammals
    B. cerebral nuclei
    C. gyri
    D. dendrites

    Ans. C

17. The protecting membranes around the brain are:

    A. cerebrospinal fluid
    B. meninges
    C. white matter
    D. cranium

    Ans. B

18. Sensory messages are transmitted to the cerebrum through what type of tracts?

    A. descending
    B. ascending
    C. commisures
    D. reflex

    Ans. B

19. The brain is a very metabolically active organ. It receives _________ of the blood pumped by the heart, and _________ of the oxygen used by the body.

    A. 2%, 10%
    B. 50%, 50%
    C. 20%, 2%
    D. 20%, 20%

    Ans. D

20. Which of the following is NOT a lobe of the cerebral hemisphere?

    A. frontal
    B. cerebral
    C. occipital
    D. parietal

    Ans. B

21. The primary motor areas controlling skeletal muscles are centered in the:

    A. parietal lobe
    B. frontal lobe
    C. temporal lobe
    D. occipital lobe

    Ans. B

22. What part of the brain maintains consciousness?

    A. limbic system
    B. cerebellum
    C. reticular activating system
    D. pons

    Ans. C

23. What part of the brain affects the emotional aspects of behaviors

    A. reticular activating system
    B. limbic system
    C. parasympathetic system
    D. peripheral nervous system

    Ans. B

24. An electroencephalogram is a record of electrical activity of the brain. Which of the following type of brain waves and their source is correctly matched?

    A. alpha waves--large wave generated at stages of sleep
    B. alpha waves--come from usual area of brain
    C. beta waves--short waves generated at stages of sleep
    D. delta waves--reflect heightened mental activity

    Ans. B

Responsiveness: Nervous Systems

25. REM sleep is characterized by periods of:

    A. increase in alpha waves
    B. rapid eye movement
    C. normal sleep
    D. lowered metabolic activity

    Ans. B

26. Information the brain is aware of at a given moment is called:

    A. sensory memory
    B. short-term memory
    C. long-term memory
    D. association memory

    Ans. B

27. Recognition and interpretation of words is permitted by the:

    A. hippocampus
    B. general interpretive area
    C. Wernicke's area
    D. auditory area

    Ans. C

28. Environmental stimulation of the brain is important

    A. only during the period of development of a child's brain
    B. only in adulthood to maintain the status of the cerebral cortex
    C. in all stages of life to develop and maintain the status of the cerebral cortex
    D. only to a limited degree because of its correlation to hereditary problems

    Ans. C

29. Which terms best contrast the somatic and autonomic nervous systems?

    A. external environment; internal environment
    B. sensory; motor
    C. organs; body
    D. peripheral; central

    Ans. A

30. Each spinal nerve has:

A. gray and white matter
B. a dorsal and a ventral root
C. ventral ganglia
D. dorsal root motor fibers

Ans. B

31. Which of the following best contrasts the sympathetic and parasympathetic systems of the autonomic nervous system?

A. mobilizes energy, conserves energy
B. sensory, motor
C. visceral, somatic
D. afferent, efferent

Ans. A

32. The parasympathetic system controls the heart by:

A. increasing the amount of blood pumped
B. increasing the blood pressure
C. slowing the heart rate
D. increasing the heart rate

Ans. C

33. Drug tolerance means:

A. drug has no negative effect on user
B. the drug reduces the body's normal production of drugs
C. more of the drug is required for desired effect
D. the drug has no effect on the individual and is safe

Ans. C

34. The majority of the brain cells are located in the:

A. medulla
B. cerebrum
C. cerebellum
D. thalamus

Ans. B

Responsiveness: Nervous Systems

35. The part of the brain continuous with the spinal cord is the:

    A. cerebellum
    B. pons
    C. medulla oblongata
    D. cerebrum

    Ans. C

36. The fundamental organization of the cerebral cortex is divided into:

    A. gray and white matter
    B. gyri and sulci
    C. right and left hemispheres
    D. motor, sensory, and associated cortex

    Ans. D

37. How do drugs alter the mood of an individual using the drug? (Give an example to demonstrate your answer)

38. How are the body organs controlled in vegetative and emergency conditions? Use the heart as an example to demonstrate your answer.

39. Hypothesize the path of sensory input and motor output to and from the brain.

40. Hypothesize why general sensory input to the cerebrum is gone when you sleep.

Chapter 35

Sensory Perception

1. The human taste bud is a sensory receptor which is a modified:

   A. nerve cell
   B. epithelial cell
   C. nerve cell ending
   D. nerve fiber

   Ans. B

2. Mechanoreceptors do NOT respond to:

   A. pressure
   B. gravity
   C. movement
   D. heat

   Ans. D

3. Which of the following classes of receptors best describes a receptor receiving stimuli from the outside environment?

   A. exteroceptor
   B. mechanoreceptor
   C. photoreceptor
   D. interceptor

   Ans. A

4. An interoceptor which detects changes in pH of body fluid is:

   A. mechanoreceptor
   B. exteroceptor
   C. chemoreceptor
   D. proprioceptor

   Ans. C

Sensory Perception

5. Sensation takes place:

   A. at the receptor
   B. in the brain
   C. in, or on the part of the body stimulated
   D. in the neuron

   Ans. B

6. In a flower shop the smell of roses disappears rapidly after you enter the shop because:

   A. ventilation removes the scents from the room
   B. of sensory adaptation
   C. floral scents stop interacting with the receptors
   D. receptors continue to respond to stimuli but the neuron does not pick up stimulus

   Ans. B

7. Later line organ in the fish is what type of receptor?

   A. thermoreceptor
   B. photoreceptor
   C. mechanoreceptor
   D. chemoreceptor

   Ans. C

8. Movement of connective tissue in a Pacinian corpuscle generates an action potential by:

   A. bending a hair
   B. vibrating a fluid
   C. stretching a muscle spindle
   D. deforming an axon

   Ans. D

9. The electrical potential which is on a receptor at all times before a stimulus arrives is called:

   A. action potential
   B. receptor potential
   C. electrical potential
   D. postsynaptic potential

   Ans. B

10. A receptor is a transducer because it converts:

    A. electrical energy to kinetic energy
    B. potential energy to kinetic energy
    C. electrical energy to a sensation
    D. any form of energy to electrical energy

    Ans. D

11. Pigment molecules in photoreceptors absorb:

    A. mechanical energy
    B. light energy
    C. chemical energy
    D. thermal energy

    Ans. B

12. Arrange the following in order of receptor activation:
    1. action potential
    2. stimulus
    3. receptor potential
    4. transduction

    A. 2,4,3,1
    B. 1,2,3,4
    C. 4,3,2,1
    D. 2,4,1,3

    Ans. A

13. A guarded response and the all-or-none principle contrasts:

    A. sensation and transduction
    B. receptor potential and an action potential
    C. receptor and stimulus
    D. electrical energy and mechanical energy

    Ans. B

14. How does the nervous system differentiate light energy, mechanical energy, and thermal energy when the action potentials are similar?

    A. each action potential is different and the brain interprets the signal
    B. the brain "knows" different energies
    C. the brain translates the quality of the potential
    D. each receptor is connected to a specific part of the brain

    Ans. D

Sensory Perception

15. Muscle spindles, golgi tendon organs, and joint receptors are classified as what type of receptor?

    A. chemoreceptor
    B. photoreceptor
    C. proprioceptor
    D. exteroceptor

    Ans. C

16. What type of receptor helps you put food in your mouth which you almost never miss, and you can not see it?

    A. proprioceptor
    B. chemoreceptor
    C. exteroceptor
    D. photoreceptor

    Ans. A

17. The part of the inner ear which is responsible for hearing is called the:

    A. saccula
    B. ventricle
    C. cackler
    D. semicircular canal

    Ans. C

18. The small calcium carbonate crystals at the inner ear which play a role in detecting gravity are called:

    A. autoliths
    B. cupula
    C. endolymph
    D. empulla

    Ans. A

19. The fluid in the semicircular canals which assists in detecting the movement of the head is called:

    A. water
    B. endolymph
    C. cerebral spinal fluid
    D. perilymph

    Ans. B

20. The receptor cell in the inner ear which makes it possible to hear and maintain equilibrium is called a(n):

    A. ampulla
    B. crista
    C. cupulae
    D. hair cell

    Ans. D

21. The vestibular canal and the tympanic canal are part of the:

    A. saccula
    B. semicircular canals
    C. cackler
    D. utricle

    Ans. C

22. The auditory receptor is

    A. crista ampullaris
    B. organ of corti
    C. vestibular apparatus
    D. labyrinth

    Ans. B

23. Sound vibrations in the air are transmitted to the cochlea through the following structures. Arrange them in the proper sequence
    1. hammer
    2. anvil
    3. ear drum
    4. stirrup

    A. 3,1,4,2
    B. 1,2,4,3
    C. 4,2,1,3
    D. 3,1,2,4

    Ans. D

Sensory Perception

24. Arrange the events involved in hearing in the proper order starting with sound waves entering the auditory canal
    1. Middle ear bones vibrate
    2. Cochlear nerve transmits impulse to brain
    3. Vibrations conducted through fluid
    4. Basila-membrane vibrates
    5. Tympanic membrane vibrates

    A. 5,1,4,3,2
    B. 5,1,3,4,2
    C. 1,5,3,4,2
    D. 5,1,3,2,4

    Ans. B

25. Input from chemoreceptors (taste buds and olfactory hairs) are interpreted in the brain to produce the:

    A. flavor of food
    B. basic tastes: sweet, bitter, sour, and salty
    C. smell of a rose
    D. taste of sugar

    Ans. A

26. Rattlesnakes use thermoreceptors to:

    A. keep their body temperature constant
    B. find their food
    C. detect cold environmental temperatures
    D. control heat generation mechanisms

    Ans. B

27. Predatory fish have receptors which can detect electrical fields generated by hearts and other organs of prey fish. Name the receptor

    A. electrical organ
    B. special olfactory receptors
    C. chemoreceptors
    D. electroreceptors

    Ans. D

28. An ocelli is a light-sensitive organ found in:

A. fish
B. mammals
C. flatworms
D. earthworms

Ans. C

29. The eye of an insect is composed of units called:

A. cornea
B. lens
C. ommatidium
D. compound eye

Ans. C

30. The outer layer of the wall of the eye ball of mammals is called:

A. choroid layer
B. sclera
C. cornea
D. vitreous body

Ans. B

31. The aqueous fluid and vitreous body give shape to the eye ball.

A. aqueous fluid is in front of the lens
B. aqueous fluid is behind the lens
C. vitreous body is in front of the lens
D. vitreous body is behind the choroid

Ans. A

32. Myopia (nearsightedness) is corrected with

A. convex lens which brings light rays to focus at a point further back
B. concave lens which brings light rays to focus at a point further forward
C. convex lens which brings light rays to focus on the retina
D. concave lens which brings light rays to focus at a point further back

Ans. D

Sensory Perception

33. Accommodation involves a change in the:

    A. pupil
    B. lens
    C. retina
    D. sclera

    Ans. B

34. Refraction of light is a function of the:

    A. lens
    B. iris
    C. pupils
    D. choroid layer

    Ans. A

35. When light enters the eye, it passes through the:
    1. lens
    2. cornea
    3. aqueous humor
    4. vitreous body
    Arrange each structure listed above in order of the passage of light into the eye:

    A. 2,1,3,4
    B. 1,2,3,4
    C. 2,3,4,1
    D. 1 2,4,3

    Ans. A

36. Rhodopsin is the visual pigment in the rods and is formed from:

    A. vitamin C
    B. vitamin A
    C. vitamin D
    D. vitamin B

    Ans. B

37. How is the mammalian eye analogous to a camera?

38. How can an individual with an amputated leg experience an itching sensation on the foot of the amputated leg?

39. If a blind rattlesnake is presented with two light bulbs and one bulb is illuminated and the other is not illuminated, which bulb will the snake strike and why?

40. Why are individuals suffering from night blindness treated with Vitamin A?

Chapter 36

Internal Transport

1. During aerobic activity the maximum heart rate for a person 35 years old should be:

   A. 125 beats per minute
   B. 165 beats per minute
   C. 185 beats per minute
   D. 220 beats per minute

   Ans. C

2. Very small organisms acquire nutrients and oxygen and dispose of metabolic waste products by what method?

   A. open circulatory system
   B. diffusion
   C. closed circulatory system
   D. partial circulatory system

   Ans. B

3. Blood or a fluid connective tissue, a heart or pumping device, and a system of blood vessels or spaces through which the blood circulates, are all components of:

   A. a closed circulatory system
   B. a partial circulatory system
   C. an open circulatory system
   D. any general circulatory system found in all animals

   Ans. A

4. Some organisms, such as cnidarians, and many flatworms combine digestive and some internal transport functions into:

   A. cardio-digestive system
   B. nutriovascular cavity
   C. cardio-nutrient system
   D. gastrovascular cavity

   Ans. D

Internal Transport

5. The circulatory system of arthropods and mollusks is to the circulatory system of annelids and echinoderms, as:

   A. parital circulatory system is to open circulatory system
   B. open circulatory system is to closed circulatory system
   C. closed circulatory system is to open circulatory system
   D. closed circulatory system is to partial circulatory system

   Ans. B

6. Capillaries are to hemocoel, as:

   A. open circulatory system is to closed circulatory system
   B. closed circulatory system is to partial circulatory system
   C. closed cirulatory system is to open circulatory system
   D. open circulatory system is to partial circulatory system

   Ans. C

7. A blood pigment that transport oxygen in some invertebrates is to the oxygen-transporting mechanism found in insects as:

   A. hemoglobulin is to lungs
   B. plasma is to hemocyanin
   C. hemocyanin is to lungs
   D. hemocyanin is to tracheal tubes

   Ans. D

8. The position of the heart in all vertebrates from fishes to the mammals is located:

   A. ventrally
   B. dorsally
   C. laterally
   D. proximally

   Ans. A

9. In vertebrates, the pale yellow fluid is to the blood cells that transport oxygen, as:

   A. urine is to red blood cells
   B. plasma is to red blood cells
   C. hemoglobin is to white blood cells
   D. urine is to platelets

   Ans. B

10. Transport substances which transport specific hormones are to proteins involved in the clotting of blood, as:

    A. gamma globulins are to albumins
    B. albumins are to platelets
    C. globulins are to serum
    D. albumins are to fibrinogen

    Ans. D

11. Antibodies that provide immunity against invading disease organisms are to the remaining liquid that is formed when the protein involved in blood clotting has been removed from the liquid portion of the blood, as:

    A. albumins are to plasma
    B. gamma globulins are to plasma
    C. albumins are to fibrinogen
    D. gamma globulins are to serum

    Ans. C

12. Red blood cells are to white blood cells, as:

    A. erythrocytes are to platelets
    B. erythrocytes are to leukocytes
    C. neutrophils are to basophils
    D. eosinophils are to basophils

    Ans. B

13. A deficiency of hemoglobin usually accompanied by a reduced number of red blood cells is to an anticlotting chemical that may be important in preventing inappropriate clotting in blood vessels, as:

    A. anemia is to histomine
    B. hemophilia is to histomine
    C. anemia is to heparin
    D. hemophilia is to agranular leukocytes

    Ans. C

14. What cellular organelle is lacking from the mature red blood cells of mammals?

    A. nucleus
    B. ribosomes
    C. mitochondria
    D. endoplasmic reticulum

    Ans. A

Internal Transport

15. Small fragments of cytoplasm that separate from certain large cells in the bone marrow, are referred to as:

    A. leukocytes
    B. erythrocytes
    C. monocytes
    D. platelets

    Ans. D

16. A plasma protein manufactured in the liver, that requires vitamin K for its production is to the insoluble protein that is converted from the soluble plasma protein fibrinogen, as:

    A. thrombin is to heparin
    B. prothrombin is to fibrin
    C. prothrombin is to histamine
    D. fibrinogen is to fibrin

    Ans. B

17. Carries blood away from the heart towards other tissues is to carries blood back toward the heart, as:

    A. veins are to arteries
    B. veins are to capillaries
    C. arteries are to veins
    D. arteries are to arteries

    Ans. C

18. A reduction in the diameter of the smooth muscles in arteries is to smooth muscle arteriole relaxation, as:

    A. vasoconstriction is to vasodilation
    B. vasodilation is to vasocontraction
    C. vasoconstriction is to vaso-release
    D. vaso-reduction is to vaso-release

    Ans. A

19. The membrane that surrounds the entire heart is to the chamber that receives blood, as:

    A. interatrial septum is to ventricle
    B. pericardium is to atrium
    C. pericardial sac is to ventricle
    D. pericardial sac is to atrioventricular

    Ans. B

20. To prevent blood from flowing backward, from the left ventricle to the left atrium, the valve that prevents this is referred to as:

A. left semilunar valve
B. tricuspid valve
C. pericardial valve
D. mitral valve

Ans. D

21. What is the name of the valves that prevent backflow of blood from the great arteries into the ventricles?

A. atrioventricular valves
B. semilunar valves
C. bicuspid valves
D. mitral valves

Ans. B

22. Cardia muscle fibers are separated at their ends by dense bands, referred to as:

A. tight junction
B. plasmids
C. intercalated discs
D. myolin sheath

Ans. C

23. The heartbeat begins in a node of specialized muscle called the pacemaker, which is also referred to as:

A. sinoatrial node
B. atrioventricular node
C. sinoventricular node
D. intercalated junction

Ans. A

24. The contraction phase of the heart is to the relaxation phase of the heart, as:

A. diastole is to systole
B. systole is to arteriostole
C. ventricularstole is to arteriostole
D. systole is to diastole

Ans. D

Internal Transport

25. The cause for an increase in the strength of a ventricular contraction is to the cause of a decrease in the strength of a ventricular contraction, as:

    A. vagus nerve is to central nervous system
    B. sympathetic nerves are to parasympathetic nerves
    C. epinephrine is to norepinephrine
    D. adrenalin is to norepinephrine

    Ans. B

26. What is the normal heart rate per minute, and volume of blood that is pumped by one ventricle per minute?

    A. 70 beats per minute, and 5 liters per minute
    B. 60 beats per minute, and 2 liters per minute
    C. 72 beats per minute, and 10 liters per minute
    D. 72 beats per minute and one liter per minute

    Ans. A

27. When a heart valve does not close properly, some blood may flow backwards, creating a hissing sound, which is referred to as:

    A. heart attack
    B. stroke
    C. murmur
    D. dysfunctional heart

    Ans. C

28. Normal blood pressure is to the possible condition caused by obesity and high dietary salt intake, as:

    A. 80/120 is to hypotension
    B. 120/200 is to hypotension
    C. 120/60 is to hypotension
    D. 120/80 is to hypertension

    Ans. D

29. Flaplike valves associated with blood vessels are to hemorrhoids, as:

    A. veins are to varicose veins
    B. arteries are to arteriosclerosis
    C. capillaries are to coagulation
    D. arteries are to angiotensin

    Ans. A

30. Circulation to the lungs is to circulation to all tissues and organs of the body, as:

    A. ventricular circulation is to atrio circulation
    B. pulmonary circulation is to systemic circulation
    C. pulmonary circulation is to somatic circulation
    D. systemic circulation is to atrioventricular circulation

    Ans. B

31. Oxygen-poor blood is to oxygen-rich blood, as:

    A. left atrium is to right atrium
    B. aorta is to pulmonary artery
    C. right atrium is to left atrium
    D. aorta is to carotid arteries

    Ans. C

32. The blood vessel the coronary arteries originate from is to the large vein that receives blood from coronary veins, as:

    A. carotid arteries are to jugular vein
    B. aorta is to coronary sinus
    C. mesenteric arteries are to subclavian veins
    D. carotid arteries are to inferior vena cava

    Ans. B

33. Normal circulation of blood is to hepatic portal circulation, as:

    A. artery to vein to capillary is to artery to tiny blood vessel to capillary to vein
    B. capillary to artery to vein is to capillary to artery to vein to tiny blood vessels
    C. capillary to vein to artery to capillary is to tiny blood vessels to vein to artery to vein
    D. artery to capillary to vein is to artery to capillary to vein to tiny blood vessels to vein

    Ans. D

34. To collect and return tissue fluid to the blood, defend the body against disease organisms, and to absorb lipids from the digestive system are all functions of:

    A. the lymphatic system
    B. the circulatory system
    C. the digestive system
    D. the spleen

    Ans. A

Internal Transport

35. What specific structures help protect the respiratory system from infection by destroying bacteria and other foreign matter that enters the body through the mouth and nose?

    A. lymph nodes
    B. interstitial fluid
    C. tonsils
    D. vomerine

    Ans. C

36. The main force that pushes plasma out of the blood is to the fluid that leaves the blood vessels or capillaries under pressure, as:

    A. potential fluid pressure is to serum fluid
    B. hydrostatic pressure is to interstitial fluid
    C. osmotic pressure is to venous fluid
    D. osmotic potential pressure is to intercapillary-cellular fluid

    Ans. B

37. Explain the functions of:
    a. platelets
    b. red blood cells
    c. monocytes
    d. macrophages
    e. neutrophils

38. Explain how the lymphatic system helps maintain fluid balance.

39. Explain the difference between pulmonary circulation and systemic circulation. Trace the blood from the right atrium to the aorta.

40. Explain how the heartbeat is initiated and regulated.

Chapter 37

Internal Defense: Immunity

1. A disease-causing organism is known as a:

   A. immunogen
   B. pathogen
   C. carcinogen
   D. mutagen

   Ans. B

2. A substance that can cause an immune response is called a(n):

   A. antibody
   B. immunoglobulin
   C. antigen
   D. pathogen

   Ans. C

3. The study of our body's defense mechanisms is the science known as:

   A. pathology
   B. bacteriology
   C. mycology
   D. immunology

   Ans. D

4. An immune system that acts as a specific defense mechanism is a/an:

   A. antibody
   B. antigen
   C. allergen
   D. pathogen

   Ans. A

Internal Defense: Immunity

5. Most invertebrates' immune systems are best described as:

   A. specific
   B. nonspecific
   C. specific and nonspecific in nature
   D. there is no immune system in invertebrates

   Ans. B

6. Which of these is a part of the vertebrate specific defense mechanisms?

   A. antibodies
   B. antlgens
   C. interferon
   D. inflammation

   Ans. A

7. Your "first line of defense" against disease is/are your:

   A. skin
   B. antibodies
   C. mucous membranes
   D. phagocytes

   Ans. A

8. A protective enzyme found in tears, sweat, saliva, and mucous secretions that helps protect you against disease-causing microorganisms is:

   A. amylase
   B. proteinase
   C. lysosomes
   D. pepsin

   Ans. C

9. When you ingest microorganisms, you are (somewhat) protected from the possibility of them causing disease by:

   A. the mucous membrane of the stomach
   B. acid secreted by the stomach
   C. antibodies secreted by the stomach
   D. antigens in the stomach

   Ans. B

10. A nonspecific protein which is secreted by immune system cells in response to infection by bruises or other intracellular parasites, and which acts to inhibit viral replication, is called a/an:

    A. lysosome
    B. interferon
    C. immunoglobulin
    D. interleukin

    Ans. B

11. Signs of localized infection DO NOT include:

    A. redness
    B. swelling
    C. fever
    D. pain

    Ans. C

12. Fever is caused by the release of a protein called __________ from immune system cells.

    A. lysosome
    B. interleukin-I
    C. interferon
    D. complement

    Ans. C

13. Fever can decrease the amount of __________ in the blood, thus inhibiting bacterial growth.

    A. calcium
    B. zinc
    C. magnesium
    D. iron

    Ans. D

14. Edema, a sign of inflammation, is:

    A. swelling due to fluid accumulation
    B. pain due to injury and pressure
    C. heat due to increased blood flow
    D. redness due to vasodilation

    Ans. A

Internal Defense: Immunity

15. A cell in the immune system that engulfs and destroys foreign material is known as a:

    A. lipocyte
    B. erythrocyte
    C. phagocyte
    D. microcyte

    Ans. C

16. A phagocytic cell that can eat as many as 100 bacteria in its lifetime, and "wanders" through tissues seeking out and destroying foreign materials is a:

    A. B-cell
    B. T-cell
    C. macrophage
    D. lymphocyte

    Ans. B

17. One advantage of having a SPECIFIC defense system is that it:

    A. has a memory against previously encountered antigens
    B. can begin working in just a few minutes after infection
    C. works only against viruses
    D. produces antigens

    Ans. A

18. Immune system cells that are responsible for making soluble antibodies are:

    A. cytotoxic T-cells
    B. helper T-cells
    C. suppressor T-cells
    D. B-cells

    Ans. A

19. Immune system cells that recognize and destroy foreign cells are:

    A. cytotoxic T-cells
    B. helper T-cells
    C. suppressor T-cells
    D. B-cells

    Ans. A

20. Immune system cells that inhibit the activity of other (immune system) cells by releasing cytokines are:

A. cytotoxic T-cells
B. helper T-cells
C. suppressor T-cells
D. B-cells

Ans. C

21. Immune system cells that activate immune response are:

A. cytotoxic T-cells
B. helper T-cells
C. suppressor T-cells
D. B-cells

Ans. B

22. The thymus gland is responsible for directing the development of:

A. macrophages
B. leukocytes
C. T-cells
D. B-cells

Ans. C

23. An immune system cell that can become an antigen-presenting cell (APC), often phagocytizing and digesting microorganisms, and then displaying antigens on its surface, is a/an:

A. macrophage
B. basophil
C. neutrophil
D. eosinophil

Ans. A

24. The major histocompatibility complex (MHC) proteins:

A. are found only on lymphocytes
B. are the same things as antibodies
C. block your ability to destroy bacteria
D. allow your immune system to "see" foreign materials as non-self

Ans. D

Internal Defense: Immunity

25. Individuals who would have EXACTLY the same MHC proteins are:

    A. all brothers and sisters of the same family
    B. identical siblings (twins, triplets,...)
    C. parents of a family
    D. tissue donors and transplant recipients

    Ans. B

26. When a B-cell comes in contact with a foreign particle, it will divide and differentiate into what types of cells?

    A. plasma cells and T-cells
    B. T-cells and memory cells
    C. memory cells and plasma cells
    D. T-cells and other B-cells

    Ans. C

27. An antigenic determinant is:

    A. a small part of an antigen that is recognized by an antibody
    B. part of an antibody molecule
    C. found on the surface of B-cells
    D. produced by T-cells

    Ans. A

28. A typical antibody molecule is shaped like the letter

    A. Z
    B. Y
    C. X
    D. V

    Ans. B

29. There are _________ classes of antibodies.

    A. three
    B. five
    C. two
    D. four

    Ans. B

30. The antibody molecule that is most prevalent in the blood of a healthy person, and also crosses the placenta from mother to fetus to confer some immunity to the infant at birth, is:

    A. IgA
    B. IgD
    C. IgG
    D. IgM

    Ans. C

31. The antibody molecules that stimulate allergies by stimulating the release of histamines is

    A. IgA
    B. IgD
    C. IgM
    D. IgE

    Ans. D

32. The antibody molecule that is found in tears, saliva, breast milk, and mucous secretions, is:

    A. IgA
    B. IgD
    C. IgG
    D. IgM

    Ans. A

33. Cell-mediated immunity uses which type of immune cell?

    A. basophils
    B. B-cells
    C. T-cells
    D. neutrophils

    Ans. C

34. Booster vaccines are given to stimulate a ________ response in the immune system.

    A. primary
    B. secondary
    C. slower
    D. broad

    Ans. B

Internal Defense: Immunity

35. Tissue transplants are "rejected" by the recipient's immune system when:

    A. B-cells recognize and lyse the foreign cells
    B. T-cells recognize foreign MHC molecules
    C. neutrophils make antibodies against the transplanted tissue
    D. tissue from an identical twin is used

    Ans. B

36. An example of an autoimmune disease, where the immune system attacks and destroys your <u>OWN</u> tissues, would be:

    A. cystic fibrosis
    B. congestive heart failure
    C. multiple sclerosis
    D. breast cancer

    Ans. C

37. A substance that causes an allergic reaction in some people is a:

    A. carcinogen
    B. allergen
    C. mutagen
    D. pathogen

    Ans. B

38. The most dangerous (possible) result of exposure to a substance you are allergic to might be:

    A. hives
    B. sensitization
    C. allergic asthma
    D. systemic anaphylaxis

    Ans. D

39. The AIDS virus most often infects:

    A. helper T-cells
    B. cytotoxic T-cells
    C. suppressor T-cells
    D. B-cells

    Ans. A

40. The drug, AZT, is useful in the treatment of AIDS because it:

A. destroys HIV
B. prevents infection of cells by HIV
C. blocks replication of HIV
D. stimulates the production of antibodies against HIV

Ans. C

41. If you have the measles or the chicken pox once, you'll probably be immune to them for life. It is very probable that if you get the "flu" one year, you can still get the "flu" again the next year (The same thing can be said about colds). Explain why you can develop life-time immunity to some diseases but (seemingly) no immunity to others.

42. There are two basic methods for acquiring immunity to certain pathogens, namely active and passive immunity. Discuss both classes of immunity, and give examples of how each type of immunity might be acquired.

43. Describe some of the ways your immune system reacts against cancer cells, and discuss how many cancers simply defeat your immune system and cause you to develop cancer anyway.

44. HIV (Human Immunodeficiency Virus) is known to cause AIDS. Discuss how HIV interacts with the immune system, and how HIV can eventually cause failure of the immune system.

Chapter 38

Gas Exchange

1. Oxygen moves into animals and carbon dioxide moves out of animals by:

   A. active transport
   B. osmosis
   C. diffusion
   D. dialysis

   Ans. C

2. In all aquatlc and terrestrlal animals gas exchange occurs:

   A. across a moist membrane surface
   B. in the blood
   C. ln a lung
   D. in a trachea tube

   Ans. A

3. The oxygen all terrestrial and aquatic animals use must:

   A. be the oxygen in a water molecule
   B. be dissolved in water
   C. be moved against a concentration gradient
   D. be exchanged for CO_2

   Ans. B

4. Which of the following does not contribute to gas exchange with the body surface?

   A. a thick dry cuticle
   B. high surface-to-volume ratio
   C. low metabolic rate
   D. mucus produced on the surface of the body

   Ans. A

Gas Exchange

5. Branching tubes that deliver air to all parts of the body of an insect are called:

   A. spiracle tubes
   B. bronchial tubes
   C. tracheal tubes
   D. parabronchi

   Ans. C

6. Gills in vertebrates are found internally and separated from the environment by:

   A. spiracles
   B. operculum
   C. dermal gills
   D. cilia on the exposed surface

   Ans. B

7. Counter current flow increases the amount of oxygen that enters the blood of birds. Which of the following is a counter-current system?

   A. oxygen and carbon dioxide diffuse in opposite directions
   B. blood flow in lungs is opposite to the flow in the parabronchi
   C. oxygen diffused against a concentration gradient
   D. air moving in wlth inhalation and out on exhalation

   Ans. B

8. The flap of tissue that closes the trachea when swallowing is the:

   A. trachea
   B. glottis
   C. epiglottis
   D. bronchi

   Ans. C

9. Air is carried into the lungs by the:

   A. trachea
   B. bronchi
   C. bronchioles
   D. epiglottis

   Ans. B

10. Arrange the following in order of the passage of air INTO the lungs.
    1. bronchus
    2. pharynx
    3. trachea
    4. larynx

    A. 2,4,1,3
    B. 2,3,4,1
    C. 2,3,1,4
    D. 2,4,3,1

    Ans. D

11. Gas exchange occurs in the:

    A. alveolus
    B. larynx
    C. pharynx
    D. bronchus

    Ans. A

12. Before oxygen gets into the blood it diffuses through:

    A. pleural membrane
    B. alveolus
    C. epithelium of the alveolar wall and walls of the capillary
    D. epithelium of the alveolus

    Ans. C

13. Relaxing the muscles of the chest wall and diaphragm:

    A. causes exhalation
    B. causes inhalation
    C. increases the volume of the chest cavity
    D. increases the pleural cavity

    Ans. A

14. The primary respiratory centers are found in the:

    A. medulla
    B. pons
    C. cerebrum
    D. cerebellum

    Ans. A

Gas Exchange

15. When carbon dioxide increases the:

    A. pH goes down
    B. pH goes up
    C. pH stays the same
    D. blood becomes basic

    Ans. A

16. Chemoreceptors for detecting carbon dioxide are found in which of the following structures?

    A. pulmonary vein
    B. carotid artery
    C. vena cava
    D. cerebrum

    Ans. B

17. Which factors listed below does Fich's law say are involved in increasing the diffusion of a gas in the lungs?

    A. the greater the surface area
    B. the lesser the surface area
    C. the layer of molecules
    D. the lesser the difference in pressure across the membrane

    Ans. A

18. Where would you find the greatest amount of carbon dioxide?

    A. in the blood
    B. in the alveolar air
    C. in the tissues
    D. in the external air

    Ans. C

19. Where would you find the greatest amount of oxygen?

    A. in the blood
    B. in the alveolar air
    C. in the tissue
    D. in the external air

    Ans. D

20. Oxygen is stored in muscle cells as:

A. hemoglobin
B. oxyhemoglobin
C. glycogen
D. protein

Ans. B

21. Carbon dioxide reacts with water to form:

A. lactic acid
B. NaHCO
C. hemoglobin
D. oxyhemoglobin

Ans. B

22. CO_2 is transported as:

A. bicarbonate ions
B. hemoglobin molecules
C. carbonic acid
D. hemoglobin

Ans. A

23. An abnormally low pH is:

A. acidosis
B. alkalosis
C. saturation
D. disassociation

Ans. A

24. The best contrast between acidosis and alkalosis is

A. high pH, low pH
B. high hydrogen concentration, low hydrogen concentration
C. high hydrogen concentration, low pH ion concentration
D. low hydrogen concentration, high pH

Ans. B

Gas Exchange

25. An important muscle in breathing is/are:

    A. the diaphragm
    B. the abdominal muscles
    C. the muscles in the lung
    D. the pectorals muscles of the chest

    Ans. A

26. Movement of air in and out of the lungs is:

    A. physiological oxidation
    B. gas exchange
    C. ventilation
    D. respiration

    Ans. C

27. Air breathed is cleaned by:

    A. particles sticking to mucus surface
    B. flushing with water
    C. the nasal sinuses
    D. swallowing

    Ans. A

28. An enzyme required to combine water and carbon dioxide is:

    A. carbonate
    B. bicarbonate
    C. carbonic anhydrase
    D. hemoglobin

    Ans. C

29. The normal body pH is:

    A. 6.1
    B. 7.0
    C. 7.4
    D. 7.8

    Ans. C

30. Oxyhemoglobin dlsassociates in a(n):

A. acid solutlon
B. basic solution
C. neutral solution
D. enzyme action

Ans. A

31. When does alr enter the body?

A. when you open your mouth
B. when pressure in lungs is below atmospheric pressure
C. when intercostal muscles relax
D. when abdominal muscles contract

Ans. B

32. When does inhalation end?

A. when diaphragm relaxes
B. when air pressure in the lungs equals the atmospheric pressure
C. when intercostal muscles relax
D. when abdominal muscles contract

Ans. B

33. Which statement best describes the movement of oxygen?

A. it diffuses to where it is used
B. it diffuses from a high concentration to a low concentration
C. it diffuses to where it is needed
D. it is transported throughout the body

Ans. B

34. Why is blood a bright red color when you cut yourself?

A. arterial blood contains oxyhemoglobin
B. venous blood contains oxyhemoglobin
C. arterial and venous blood will form oxyhemoglobin when exposed to air
D. arterial blood contains hemoglobin

Ans. C

Gas Exchange

35. Which oxygen is used first by the cell?

    A. oxygen bound to hemoglobin
    B. oxygen dissolved in plasma
    C. oxygen in tissue fluids
    D. oxygen dissolved in cytoplasm

    Ans. D

36. How many pounds of air do you breathe each day?

    A. 10 pounds
    B. 50 pounds
    C. 100 pounds
    D. more than 6 times the amount of food and drink you consume each day

    Ans. D

37. Fish in a tank that is poorly aerated usually come to the surface and even gulp air. Why? What are the advantages a terrestrial animal has breathing when compared to fish getting oxygen from the water?

38. If there are no muscles in the lungs, how do the muscles of the rlb cage and diaphragm increase the air in the lungs? What would you propose if the chest wall were punctured in an accident? Explain your answers in terms of pressure changes.

39. When you run, you will begin to breathe faster. Explain how the physiological mechanisms that result in breathing faster.

40. Explain why mouth-to-mouth resuscitation can save the life of an individual. Answer with an explanation of the source of the oxygen.

Chapter 39

Processing Food

1. Substances present in food that are used as an energy source for an organism are known as:

   A. calories
   B. nutrients
   C. forage
   D. rations

   Ans. B

2. An example of an organism that has a single opening for its digestive system might be a/an:

   A. earthworm
   B. fish
   C. hydra
   D. fruit fly

   Ans. C

3. How are digested food materials delivered to the body cells of an organism that has only one digestive opening?

   A. by diffusion through tissue fluids
   B. through their excretory system
   C. by muscle movement
   D. through their respiratory system

   Ans. A

4. In humans, all digestive processes begin in the:

   A. stomach
   B. small intestine
   C. mouth
   D. liver

   Ans. C

Processing Food

5. The teeth that are best suited for tearing food, especially meats, are the:

   A. incisors
   B. molars
   C. premolars
   D. canines

   Ans. D

6. The soft inside area of a tooth where nerves and blood and lymph vessels are normally found is called the:

   A. dentin
   B. enamel
   C. pulp cavity
   D. gingiva

   Ans. C

7. The enzyme that is secreted with saliva into the mouth to initiate carbohydrate digestion is called (salivary):

   A. cellulase
   B. amylase
   C. nuclease
   D. lipase

   Ans. B

8. Peristalsis is:

   A. wave-like contractions of muscles in the digestive tract
   B. the digestion of food for its nutritional content
   C. removal of waste products by the excretory system
   D. the loss of nutrients due to inefficiency in digestion

   Ans. A

9. The esophagus is a muscular tube that connects the:

   A. stomach and liver
   B. small intestine and stomach
   C. mouth and stomach
   D. small intestine and liver

   Ans. C

10. The enzyme found in the stomach that initiates protein digestion is:

    A. chymotrypsin
    B. pepsin
    C. papain
    D. glucagon

    Ans. B

11. The "valve" that controls the flow of partially-digested material from the stomach into the small intestine is the:

    A. esophageal sphincter
    B. ileocaecal sphincter
    C. anal sphincter
    D. pyloric sphincter

    Ans. D

12. The pH of the stomach is so low because the gastric glands secrete:

    A. hydrochloric acid
    B. sulfuric acid
    C. phosphoric acid
    D. citric acid

    Ans. A

13. In most vertebrates, where will the most CHEMICAL digestive processes take place?

    A. mouth
    B. small intestine
    C. stomach
    D. large intestine

    Ans. B

14. The first section of the small intestine (where digestive enzymes and substances from the liver and pancreas will be added to food materials for digestion) is known as the:

    A. ileum
    B. jejunum
    C. duodenum
    D. colon

    Ans. C

Processing Food

15. What is one reason the lining of the small intestine is so extensively folded?

    A. to decrease overall length of the intestines
    B. to increase the absorptive surface area
    C. to resist disease processes
    D. to allow wastes to pass through quickly

    Ans. B

16. Bile is produced by the:

    A. liver
    B. pancreas
    C. stomach
    D. gall bladder

    Ans. A

17. Bile is used in the mechanical digestion of:

    A. proteins
    B. carbohydrates
    C. vitamins
    D. fats

    Ans. D

18. What organ stores bile until it is needed for digestive processes?

    A. pancreas
    B. large intestine
    C. stomach
    D. gall bladder

    Ans. D

19. If you drink alcoholic beverages or take certain drugs, they are most often broken down by detoxification processes by the:

    A. liver
    B. gall bladder
    C. pancreas
    D. stomach

    Ans. A

20. Deoxyribonuclease and ribonuclease are used to break down:

A. proteins
B. lipids
C. carbohydrates
D. nucleic acids

Ans. D

21. The organ that secretes digestive enzymes and exerts a hormone (insulin) to control the level of glucose in the blood is the:

A. liver
B. stomach
C. pancreas
D. gall bladder

Ans. C

22. What type of cell lines the intestinal walls?

A. epithelial
B. muscle
C. nervous
D. connective

Ans. A

23. Lipids are absorbed by the _________ system.

A. urinary
B. blood vascular
C. respiratory
D. lymphatic

Ans. D

24. Diarrhea can result from:

A. materials passing through the large intestine too slowly
B. materials passing through the small intestine too slowly
C. materials passing through the large intestine too quickly
D. materials passing through the small intestine too quickly

Ans. C

Processing Food

25. Constipation can result from:

A. materials passing through the large intestine too slowly
B. materials passing through the small intestine too slowly
C. materials passing through the large intestine too quickly
D. materials passing through the small intestine too quickly

Ans. A

26. Place the following parts of the large intestine in their proper order.

1. sigmoid colon
2. caecum
3. descending colon
4. anus
5. transverse colon
6. rectum
7. ascending colon

A. 1-3-7-5-2-4-6
B. 2-5-1-3-7-4-6
C. 1-6-7-3-5-4-2
D. 2-7-5-3-1-6-4

Ans. D

27. About 50% of the caloric intake in the diet is in the form of:

A. proteins
B. carbohydrates
C. lipids
D. nucleic acids

Ans. B

28. About 40% of the calories taken in through the diet are:

A. proteins
B. carbohydrates
C. lipids
D. nucleic acids

Ans. C

29. The intake of ________ tends to raise blood cholesterol levels.

A. monounsaturated fats
B. diunsaturated fats
C. polyunsaturated fats
D. saturated fats

Ans. D

30. We must eat proteins in order to receive what compounds to build our own proteins from?

    A. monosaccharides
    B. nucleotides
    C. amino acids
    D. fatty acids

    Ans. C

31. Organic nutrients that are needed in smaller amounts, and that are usually used as coenzymes (for proper enzyme functioning) are known as:

    A. vitamins
    B. minerals
    C. amino acids
    D. sugars

    Ans. A

32. A deficiency in which vitamin can result in "scurvy," a condition that can afflict individuals on long land or sea voyages?

    A. vitamin A
    B. vitamin C
    C. vitamin E
    D. vitamin K

    Ans. B

33. A deficiency of which vitamin can lead to a condition known as "rickets," a disease manifested as weakened bones and teeth due to poor absorption of calcium?

    A. vitamin B1
    B. vitamin D
    C. vitamin B12
    D. vitamin K

    Ans. C

34. Obesity affects nearly _________ of working adults in the U.S.

    A. one-half
    B. one-third
    C. one-fourth
    D. one-fifth

    Ans. B

Processing Food

35. Kwashiorkor, one of the world's most widespread diseases of malnutrition, results from a lack of what nutrient in the diet?

    A. carbohydrates
    B. fats
    C. vitamins
    D. proteins

    Ans. D

36. Your cold cereal box says that it is fortified with vitamins and minerals. What are minerals?

    A. carbohydrates
    B. amino acids
    C. inorganic ions and salts
    D. fatty acids

    Ans. C

37. You've just eaten a quickly-prepared protein patty on bread, deep-fat-fried potato slices, and a carbonated beverage (containing sugar and cola extract), i.e., a burger, fries, and a soda. Discuss the fate of the types of nutrients ingested. Include in your discussion where the digestion of each component of the foodstuffs starts and ends.

38. Trace a bolus of food from ingestion to excretion, describing the function of each digestive organ the material touches.

39. It's been well documented that if a person goes on a crash diet, the weight loss will be countered by a weight-gain that usually
    exceeds the initial loss. Discuss the reasons behind this phenomenon, and describe how an individual could avoid the "rebound effect" described above.

40. Some people in parts of the world suffer from malnutrition, even though they may get plenty to eat. For example, malnutrition can
    be seen in many areas where rice is eaten as a staple in the diet (and "everybody knows" that rice is high in protein). Why might you think malnutrition is still a serious threat in an area such as this?

41. High levels of _________ have been directly associated with heart disease from atherosclerosis.

A. high-density lipoproteins
B. low-density lipoproteins
C. polyunsaturated fatty acids
D. complex amino acids

Ans. B

42. A person on a strict vegetarian diet must pay close attention to the amounts and types of what nutrients in the plants that make up their diet?

A. proteins
B. carbohydrates
C. fats
D. minerals

Ans. A

43. A vitamin that is produced by symbiotic bacteria living in the large intestine, and which is absorbed from there to eventually help with blood clotting, is:

A. vitamin A
B. vitamin D
C. vitamin C
D. vitamin K

Ans. D

44. Fluoride, a mineral often added to municipal water systems, is important in:

A. calcium absorption and deposition
B. maintenance of fluid balance
C. component of hemoglobin
D. resistance of teeth to decay

Ans. D

45. Iodine, which can be found added to most table salt in small amounts, is important for:

A. muscle contractions
B. transmission of nerve impulses
C. making thyroid hormones, which regulate metabolism
D. normal hemoglobin synthesis

Ans. C

Chapter 40

Osmoregulation, Disposal of Metabolic Wastes, Temperature

1. Which process regulates the osmotic pressure of body fluids so they do not become too dilute or too concentrated?

   A. excretion
   B. osmoregulation
   C. osmosis
   D. dialysis

   Ans. B

2. What is the best contrast of excretion and elimination?

   A. osmosis, dialysis
   B. absorbed, not absorbed
   C. nutrient, waste
   D. digested, undigested

   Ans. B

3. What process rids the body of excess water, salts, metabolic wastes, and toxins?

   A. dialysis
   B. osmoregulation
   C. absorption
   D. excretion

   Ans. D

4. Which of the following wastes is disposed of by system other than the kidney?

   A. water
   B. nitrogenous wastes
   C. carbon dioxide
   D. salts

   Ans. C

Osmoregulation, Disposal of Metabolic Wastes, Temperature

5. The nitrogenous wastes, ammonia, uric acid, and urea, are formed when what group of molecules are deaminated?

   A. bicarbonate ions
   B. fatty acids
   C. simple sugars
   D. amino acids

   Ans. D

6. Which organism disposes of its wastes through the production of urea?

   A. horse
   B. rattlesnake
   C. pigeon
   D. eagle

   Ans. A

7. Marine invertebrates which are in osmotic equilibrium with the surrounding sea water are known as:

   A. osmotic regulators
   B. dialysizers
   C. osmotic conformers
   D. filtration processors

   Ans. C

8. The excretory system of insects consists of which of the following?

   A. meta nephridia
   B. malpighian tubules
   C. protonephridia
   D. nephrons

   Ans. B

9. The primary osmoregulatory organ in vertebrates is the

   A. kidney
   B. skin
   C. lungs
   D. gills

   Ans. A

10. Fish and amphibians absorb water from their environment and lose sodium to the environment. How do they reclaim sodium from the environment?

A. consume salts in food
B. diffusion
C. active transport
D. facilitated transport

Ans. C

11. Sharks fight a high salt concentration in their environment by concentrating:

A. salts in the blood
B. urea in their blood
C. increased water retention
D. osmolregulation through the kidneys

Ans. B

12. The kidneys play an important role in waste disposal. What other structures dispose of a significant amount of metabolic wastes?

A. digestive tract
B. sweat glands
C. liver
D. all glands

Ans. B

13. Bile pigments produced from the breakdown of red blood cells are excreted by the liver in the:

A. sweat
B. urine
C. feces
D. mucous

Ans. C

14. The outer layer of the kidney is the:

A. renal cortex
B. renal medulla
C. renal pelvis
D. nephron

Ans. A

15. Arrange the following in the order of the direction of the flow of urine from its site of production.
    1. ureter
    2. kidney
    3. bladder
    4. urethra

A. 1,2,3,4
B. 2,1,3,4
C. 3,4,2,1
D. 4,3,2,1

Ans. B

16. Why is a bladder infection less common in males than in females?

A. males have a stronger immune system
B. females urethra is close to the vagina
C. the male urinary opening is more exposed to environment
D. males have a longer urethra

Ans. D

17. Arrange the following in order of the path of urine formation in the nephron.
    1. loop of Henle
    2. collecting duct
    3. Bowman's capsule
    4. proximal convoluted tubules
    5. distal convoluted tubules

A. 3,4,5,1,2
B. 3,4,1,5,2
C. 4,3,1,5,2
D. 2,1,3,4,5

Ans. B

18. Arrange the following in order of the movement of blood from the point of filtration of the blood.
    1. efferent arteriole
    2. capillary of glomerulus
    3. veins in the kidney
    4. secondary set of capillaries
    5. renal vein

A. 2,1,4,3,5
B. 2,1,3,4,5
C. 2,1,4,5,3
D. 1,2,3,4,5

Ans. A

19. Which of the following is NOT required for filtration?

A. great permeability of glomerular capillaries
B. large amount of metabolic wastes
C. high hydrostatic pressure in glomerulus
D. large surface area in glomerulus

Ans. B

20. _________ of cardiac output goes to the kidney. Every 24 hours _________ of filtrate is produced, resulting in the excretion of urine.

A. 50%, 35 gallons, 1 liter
B. 25%, 45 gallons, 1.5 liters
C. 20%, 10 gallons, 2 liters
D. 25%, 30 gallons, 4 liters

Ans. B

21. When the amount of a substance exceeds the renal threshold

A. the excess is lost in the urine
B. the excess will not be filtered
C. the filtration rate is slowed down
D. the filtration pressure is reduced

Ans. A

22. Which of the following is NOT a factor in urine formation?

A. reabsorption
B. filtration
C. secretion
D. dehydration

Ans. D

23. Most of the secretion occurs in the:

A. proximal convoluted tubules
B. loop of Henle
C. distal convoluted tubules
D. collecting ducts

Ans. C

Osmoregulation, Disposal of Metabolic Wastes, Temperature

24. Reabsorption in the kidney is a result of

    A. filtration, diffusion, and osmosis
    B. dialysis, filtration, and osmosis
    C. active transport, filtration, and dialysis
    D. active transport, diffusion, and osmosis

    Ans. D

25. The loop of Henle has an ascending and descending limb. The descending limb permits water to leave and salt to enter the forming urine. The ascending limb:

    A. is permeable to both salt and water
    B. is more permeable to salt and less permeable to water
    C. permits only water to be moved
    D. permits only salts to be moved

    Ans. B

26. At the bottom of the loop of Henle the filtrate has become:

    A. low in salt
    B. high in salt
    C. high in water
    D. isotonic

    Ans. B

27. When ADH is produced the urine becomes:

    A. concentrated and has small volume
    B. dilute and has large volume
    C. isotonic and has small volume
    D. concentrated and has large volume

    Ans. A

28. ADH produced by the hypothalamus controls permeability of collecting ducts, to:

    A. salt
    B. water
    C. nutrients
    D. wastes

    Ans. B

29. What controls the release of ADH from the hypothalamus?

A. volume of urine in the bladder
B. concentration of salts in the blood
C. amount of material reabsorbed
D. volume of filtrates moving through the nephron

Ans. B

30. What hormone controls the sodium concentration in the blood?

A. antidiuretic hormone
B. aldosterone
C. epinephrine
D. angiotensin

Ans. B

31. What stimulates the adrenal cortex to release aldosterone?

A. high salt content in the blood
B. low concentration of salt in the blood
C. increase in blood pressure
D. decrease in blood pressure

Ans. D

32. Arrange the following in the order which controls blood pressure
1. kidney releases renin
2. angiotensin is formed
3. blood pressure is low
4. blood pressure increases
5. blood vessels constrict

A. 3,2,1,5,4
B. 3,1,2,5,4
C. 2,3,1,5,4
D. 3,1,2,4,5

Ans. B

33. Which of the following may contribute to the color and odor of urine?

A. salt
B. urea
C. bile pigments
D. nitrogenous wastes

Ans. C

Osmoregulation, Disposal of Metabolic Wastes, Temperature

34. Ectotherms:

    A. produce heat from food consumed
    B. absorb heat from their environment
    C. regulate their body temperature
    D. have a constant body temperature

    Ans. B

35. Which of the following best contrasts ectotherms and endotherms?

    A. from within, from outside
    B. environmental control, self-controlled
    C. warm, cold
    D. mammals, fish

    Ans. B

36. Which of the following contribute to lowering body temperatures?

    A. constriction of blood vessels in the skin
    B. dilation of deep blood vessels
    C. increase in metabolism
    D. dilation of blood vessels in the skin

    Ans. D

37. When you pick up a bird, why does it not urinate like a frog does when you suddenly pick it up? Discuss the physiological and morphological significance of the observation.

38. Why do fishes produce hypotonic urine and terrestrial animals produce hypertonic urine? Discuss the physiological and environmental aspects of this observation.

39. Compare the metanepridial mechanism of waste disposal in earthworms with the peritoneal dialysis in humans when the kidneys have failed.

40. If you are in a life boat on the ocean, you should not drink sea water. Using your knowledge of how organisms handle environmental water and salt, explain why you should not drink sea water.

Chapter 41

Endocrine System

1. A ductless gland is a/an:

   A. exocrine gland
   B. endocrine gland
   C. sweat gland
   D. salivary gland

   Ans. B

2. A chemical messenger that regulates the activity of many tissues and organs in the body is:

   A. an enzyme
   B. a neurotransmitter
   C. a hormone
   D. a catalyst

   Ans. C

3. A negative feedback is a mechanism in which the hormone:

   A. has a direct enhancing effect on the stimulus
   B. does not return the stimulus to the homeostatic level
   C. has an opposite effect to the stimulus
   D. adapts to the stimulus

   Ans. C

4. The parathyroid gland produces the hormone which increases the available

   A. parathyroid hormone, calcium
   B. parathyroid hormone, iron
   C. thyroxine, calcium
   D. insulin, calcium

   Ans. A

5. How does a specific cell respond to a hormone?

   A. a cell will respond to only one hormone, when hormone is present the cell becomes active
   B. hormones change the activity of all cells
   C. the cell has a receptor protein which when exposed to specific hormones, changes the activity of the cell
   D. the phospholipids in the cell membrane respond to specific hormones by changing the activity of the cell.

   Ans. C

6. A hormone which passes through the cell membrane and combines with a cytoplastic protein receptor, changes the activity of the cell:

   A. by acitivating genes which translates into a protein which directly changes the activity of the cell
   B. by the receptor-hormone complex itself directly changing the activity of the cell
   C. by altering existing m/RNA which translates into proteins, which changes the activity of the cell
   D. by activating existing proteins in the cytoplasm which change the activity of the cell

   Ans. A

7. Hormones which can not pass through the cell membrane change the activity of the cell:

   A. by changing the cell membrane to a form which allows the hormone entrance
   B. by changing the permeability of the cell membrane and blocking the release of products of the cell
   C. by the hormone being altered by enzymes in cell membranes
   D. by activating a second messenger in the cytoplasm (cAMP) through enzymes in the membrane of the cell

   Ans. D

8. Prostaglandins are chemical mediators which

   A. are produced in one area and stored in another area
   B. are produced in a specific area and have broad general effects throughout the body
   C. are produced in a specific area and have effects on tissue in the area where they are produced
   D. are produced in an organ or gland and affect distant organs or tissues in the body

   Ans. C

9. Prostaglandins:

   A. have very limited clinical use
   B. have a potential for a wide range of clinical applications
   C. are not presently used in medical treatment
   D. are only used to treat conditions of the prostate gland

   Ans. B

10. A hormone produced by a neuron is:

   A. a prostaglandin
   B. a neurohormone
   C. an endorgan hormone
   D. a neurotransmitter

   Ans. B

11. The posterior lobe of the pituitary produces:

   A. antidiuretic hormone
   B. adrenocortictrophlc hormone
   C. prolactin
   D. growth hormone

   Ans. A

12. The thyroid gland produces a hormone which

   A. maintains sodium and phosphate balance
   B. promotes spermatogenesis
   C. lowers blood calcium levels
   D. stimulates mucus production

   Ans. C

13. The islets of Langerhans of the _________ produce a hormone ________ which lowers the blood glucose levels.

   A. pancreas, glucagon
   B. adrenal medulla, norepinephrine
   C. pancreas, insulin
   D. adrenel cortex, cortisol

   Ans. C

Endocrine System

14. The target organ for oxytocin is:

    A. kidney
    B. uterus
    C. liver
    D. ovary

    Ans. B

15. A group of steroid hormones produced by the adrenal cortex are called minerato-corticoids. They include:

    A. testosterone
    B. epinephrine
    C. glucocorticoids
    D. aldosterone

    Ans. D

16. The hormone which controls the thyroid gland from the anterior pituitary is

    A. TSH
    B. ACTH
    C. GH
    D. CH

    Ans. A

17. Follicle stimulating hormone (FSH) of the anterior pituitary stimulates the growth of the:

    A. ovary only
    B. testis only
    C. both the ovary and the testis
    D. secondary sex characteristics

    Ans. C

18. The drug Pitocin is used clinically to induce labor and mimics the effects of

    A. estrogen
    B. oxytocin
    C. progesterone
    D. cortisol

    Ans. B

19. Besides controlling labor, oxytosin is responsible for:

    A. development of the uterus
    B. ovulation
    C. milk let down
    D. menstruation

    Ans. C

20. The growth hormone is one of several factors which determine how tall you will be when you mature. Which of the following is NOT a function of the growth hormone?

    A. increase cellular uptake of amino acids
    B. increase deposition of calcium in bone
    C. mobilization of fat from adipose tissue
    D. increase free fatty acids in blood

    Ans. B

21. The growth hormone is regulated by growth hormone releasing and inhibiting hormones which are produced by the:

    A. growing bones
    B. pituitary
    C. liver
    D. hypothalamus

    Ans. D

22. What hormones are required for normal growth but are also involved in the termination of growth?

    A. sex hormones
    B. thyroid hormones
    C. insulin
    D. adrenal hormones

    Ans. A

23. Abnormally low production of the growth hormone results in:

    A. acromegaly
    B. pituitary dwarfs
    C. gigantism
    D. obesity

    Ans. B

24. What hormone is essential for a tadpole to develop into an adult frog?

    A. growth hormone
    B. insulin
    C. thyroxine
    D. calcitonin

    Ans. C

25. The thyroid stimulating hormone controls thyroid activity:

    A. through the second messenger mechanism
    B. by combining with a receptor protein in the cytoplasm of cells in the thyroid gland
    C. by inhibiting the pituitary glands production of its hormones
    D. by inhibiting cAMP

    Ans. A

26. The thyroid hormone is controlled by a negative feedback mechanism. Listed below are some events in the control of the thyroid hormone. Arrange them in proper order for negative feedback control.
    1. thyroid hormone concentration decreases
    2. hypothalamus secretes less hormone
    3. thyroid gland secretes less hormone
    4. concentration of thyroid hormones is high
    5. thyrold hormone inhibits the anterior pituitary

    A. 1,3,4,5,2
    B. 5,4,2,3,1
    C. 5,4,2,1,3
    D. 4,5,2,3,1

    Ans. D

27. What hormone requires a proper intake of iodine for normal and ongoing hormone production?

    A. calcitonin
    B. thyroxin
    C. insulin
    D. epinephrine

    Ans. B

28. The parathyroid hormone controls the level of _________ in the blood and _________ has the opposing effect.

    A. Ca, thyroxine
    B. Ca, calcitonin
    C. Fe, glucagon
    D. Zn, epinephrine

    Ans. B

29. Which of the following pairs of terms is properly matched?

    A. thyroid - glucagon
    B. islets of Langerhans - calcitonin
    C. beta cells - insulin
    D. alpha cells - insulin

    Ans. C

30. Insulin:

    A. increases the use of fat
    B. promotes the burning of amino acids
    C. lowers levels of glucose in the blood
    D. increases the release of glucose from the liver

    Ans. C

31. Glucagon:

    A. mobilizes fatty acids, amino acids, and glucose
    B lowers blood sugar
    C. converts glucose to glycogen
    D. blocks the formation of glucose from other metabolites

    Ans. A

32. Arrange the following events in order reflecting the control of blood sugar by insulin.
    1. blood sugar stimulates beta cells
    2. blood glucose decreases
    3. blood glucose level high
    4. beta cells increase the release of insulin

    A. 4,2,3,1
    B. 3,1,4,2
    C. 2,3,1,4
    D. 3,1,2,4

    Ans. B

33. Arrange the following events in order reflecting the control of blood sugar by glucagon.
    1. glucose concentration in blood is low
    2. alpha cells increase production of glucagon
    3. glucose stimulates alpha cells
    4. glucose concentration increases

    A. 1,3,2,4
    B. 1,4,2,3
    C. 1,2,3,4
    D. 4,3,2,1

    Ans. A

34. Insulin receptor responding to insulin and not responding to insulin contrasts:

    A. insulin and glucagon
    B. Type I and Type II diabetes
    C. Type II diabetes and Type I diabetes
    D. glucagon and insulin

    Ans. B

35. Hormones at the adrenal medulla cause blood to be rerouted to those organs essential for:

    A. digesting food
    B. thinking
    C. emergency action
    D. writing a letter

    Ans. C

36. Explain why the thyroid gland enlarges to a goiter when there is a shortage of iodine in the diet and why salt producers put iodine in salt.

37. Create an anology between a negative feedback mechanism and the brakes on a car.

38. If a friend had diabetes which is not under control, what are some signs you could look for that would indicate the friend is getting worse?

39. Knowing what you know about the effects of the growth hormone, what role does it play after growth stops?

Chapter 42

Reproduction

1. It's estimated that in the United States there are approximately how many teen-agers infected with a sexually transmitted disease?

   A. 10,000
   B. 100,000
   C. 1 million
   D. 10 million

   Ans. B

2. Asexual reproduction **usually** involves:

   A. one parent organism
   B. two parent organisms
   C. the formation of gametes
   D. the formation of diploid zygotes

   Ans. A

3. The type of asexual reproduction, that leads to the development of an adult organism from an unfertilized egg (as exhibited by honeybees) is known as:

   A. binary fission
   B. budding
   C. fragmentation
   D. parthogenesis

   Ans. D

4. An example of an organism that would use external fertilization as its means of sexual reproduction is a:

   A. human
   B. dog
   C. cat
   D. frog

   Ans. D

Reproduction

5. An animal species that normally exhibits hermaphroditism is:

   A. frog
   B. bird
   C. earthworm
   D. wasp

   Ans. C

6. An advantage of using sexual reproduction (instead of asexual reproduction) to propagate your species might be:

   A. all offspring that result are identical to the parent organism
   B. there is (usually) a need for only one parent organism
   C. there is a mixing of parental genes
   D. the offspring that result are haploid

   Ans. C

7. The cellular division process that forms gametes is:

   A. mitosis
   B. meiosis
   C. parthogenesis
   D. binary fission

   Ans. B

8. The process of spermatogenesis results in the formation of:

   A. two diploid sperm cells
   B. four diploid sperm cells
   C. two haploid sperm cells
   D. four diploid sperm cells

   Ans. C

9. Spermatogenesis takes place in what male structure(s)?

   A. seminiferous tubules
   B. seminal vesicles
   C. vas deferens
   D. epididymis

   Ans. A

10. The structure that is common to both the urinary and reproductive tracts in human males is the:

    A. urethra
    B. vas deferens
    C. bulbourethral glands
    D. testes

    Ans. A

11. The primary MALE sex hormone is:

    A. gonadotropin-releasing hormone
    B. follicle-stimulating hormone
    C. testosterone
    D. luteinizing hormone

    Ans. C

12. Complete the following analogy: Testes:Spermatogenesis

    A. Oviducts:Oogenesis
    B. Ovaries:Oogenesis
    C. Uterus:Fertilization
    D. Vagina:Fertilization

    Ans. B

13. The structure responsible for transporting an ovum after it has been shed at ovulation, as well as serving as the site of fertilization, is the:

    A. cervix
    B. vagina
    C. uterus
    D. oviduct

    Ans. D

14. A "Pap smear" is normally used to detect a cancer or pre-cancerous condition of the:

    A. ovaries
    B. oviducts
    C. vagina
    D. cervix

    Ans. D

Reproduction

15. The external genital structure in human females that is analogous to the penis in human males is the:

    A. mons pubis
    B. labia majora
    C. clitoris
    D. labia minora

    Ans. C

16. Oogenesis will produce:

    A. four ova
    B. three ova and one polar body
    C. two ova and two polar bodies
    D. one ovum and three polar bodies

    Ans. D

17. Complete the following analogy - Testosterone:Male as...

    A. estrogen:female
    B. progesterone:female
    C. gonadotropin-releasing hormone:female
    D. luteinizing hormone:female

    Ans. A

18. The AVERAGE length of the menstrual cycle is:

    A. 7 days
    B. 14 days
    C. 21 days
    D. 28 days

    Ans. D

19. During the AVERAGE-length menstrual cycle, ovulation will occur on or about:

    A. day 12
    B. day 14
    C. day 16
    D. day 18

    Ans. B

20. The female hormone that caused the development of ova just prior to their release is:

A. estrogen
B. progesterone
C. follicle-stimulating hormone
D. luteinizing hormone

Ans. C

21. Where would you find a corpus luteum?

A. in the uterus when a fertilized ovum implants
B. in the oviduct as an ovum is fertilized
C. at the cervix during pregnancy
D. in the ovary after ovulation

Ans. D

22. The initials **PMS** stand for:

A. premenstrual syndrome
B. postmenstrual syndrome
C. premenopausal syndrome
D. postmenopausal syndrome

Ans. A

23. Place the phases of sexual excitement in their correct order, from beginning to ending.
1. Excitement
2. Orgasm
3. Desire
4. Resolution

A. 3-1-4-2
B. 2-4-3-1
C. 3-1-2-4
D. 2-3-4-1

Ans. C

24. Ejaculation (release of sperm from the male body) will happen during which phase sexual excitement?

A. excitement
B. orgasm
C. desire
D. resolution

Ans. B

Reproduction

25. Fertilization of an egg with a sperm, and the following development of a fetus, is known as:

    A. contraception
    B. inception
    C. conception
    D. proception

    Ans. C

26. It will take a sperm approximately ________ to reach an egg after introduction into the female reproductive tract.

    A. 5 minutes
    B. 5 hours
    C. 5 days
    D. two weeks

    Ans. A

27. When does the second meiotic division occur in an egg cell?

    A. before birth
    B. shortly after birth
    C. at the time of ovulation
    D. at the time of fertilization

    Ans. D

28. How long does a sperm retain its ability to fertilize an egg cell after it has been introduced into the female reproductive tract?

    A. one day
    B. two days
    C. three days
    D. one week

    Ans. B

29. Most oral contraceptives prevent pregnancies by:

    A. preventing fertilization
    B. preventing implantation into the uterine wall
    C. preventing a fertilized egg from dividing
    D. preventing ovulation

    Ans. D

30. The contraceptive that works by being insert into the uterus and left until pregnancy is desired is a/an:

    A. I.U.D.
    B. diaphragm
    C. norplant
    D. condom

    Ans. A

31. The type of contraceptive that has been associated with potentially serious side effects when used by women who smoke cigarettes is the:

    A. diaphragm
    B. IUD
    C. oral contraceptive
    D. norplant

    Ans. C

32. The male sterilization technique of a vasectomy involves cutting and sealing of the:

    A. vasa deferentia
    B. urethra
    C. seminal vesicles
    D. seminiferous tubules

    Ans. A

33. A vasectomy may be reversed in approximately what percentage of sterilized men?

    A. 25%
    B. 50%
    C. 75%
    D. 0%

    Ans. A

34. Surgical sterilization of females is achieved by cutting and sealing:

    A. oviducts
    B. vagina
    C. cervix
    D. uterus

    Ans. A

Reproduction

35. An abortion is:

    A. surgical sterilization of a man
    B. surgical sterilization of a woman
    C. termination of a pregnancy
    D. removal of the uterus

    Ans. C

36. At the time the text was written, the most common sexually transmitted disease (STD) in the United States was:

    A. AIDS
    B. syphilis
    C. gonorrhea
    D. chlamydia

    Ans. D

37. Trace the path of a sperm cell from where it forms inside the male until it fertilizes an egg cell. Include in your description all structures, male and female, that the sperm cell must encounter.

38. There are three types of abortions (namely, Spontaneous abortion, Therapeutic abortion, and Terminating abortion). Discuss the significance of each type of abortion (e.g, what causes it or why is it done).

39. It is estimated that about 400 million sperm are released into the female reproductive tract at the time of ejaculation. Discuss possible reasons why such a vast number of sperm are produced and released if only one sperm cell will fertilize an egg.

40. Discuss the use of certain contraceptive devices and techniques such as IUD's, the "pill," a diaphragm, the "rhythm-method," and condoms. Include in your discussion both positive and negative aspects of the use of each.

41. It is estimated that approximately how many sexually active young adults in the U.S. will be infected with an STD (sexually-transmitted disease) by the age of 21?

A. 0%
B. 50%
C. 75%
D. 100%

Ans. B

42. An organism described as a hermaphrodite:

A. has only male sex organs
B. has only female sex organs
C. has both male an female sex organs
D. has no sex organs

Ans. C

43. A eunuch (or eunuchism) results from:

A. removal of the penis before puberty
B. removal of the testes before puberty
C. removal of the penis after puberty
D. removal of the testes after puberty

Ans. B

44. The two hormones commonly found in combination in oral contraceptives are:

A. estrogen and FSH
B. progesterone and LH
C. progesterone and estrogen
D. FSH and LH

Ans. C

45. The female reproductive structure that receives the penis during copulation (coitus) is the:

A. vagina
B. cervix
C. uterus
D. oviducts

Ans. A

Reproduction

46. The structure(s) also known as the fallopian tubes are the:

    A. vagina
    B. uterus
    C. ovaries
    D. oviducts

    Ans. D

47. Lactation occurs in:

    A. ovaries
    B. uterus
    C. oviducts
    D. breasts

    Ans. D

48. An advantage of breast-feeding infants (rather than bottle-feeding) is:

    A. a woman cannot become pregnant during the time a baby is breast-fed
    B. breast milk contains antibodies that help protect an infant from diseases
    C. breast-feeding causes the uterus to remain "stretched" after pregnancy
    D. there is no advantage to breast-feeding over bottle-feeding

    Ans. B

49. The structure in the male reproductive tract that serves as the site of maturation and storage of sperm is the:

    A. testes
    B. epididymis
    C. urethra
    D. vas deferens

    Ans. B

Chapter 43

Development

1. What is the name of the technique in which an ovum is fertilized with sperm in laboratory glassware?

   A. extra-cellular fertilization
   B. in vitro fertilization
   C. inter-uterine fertilization
   D. extraovarian conception

   Ans. B

2. A major cause of male infertility is to scarring of the oviducts in females, as:

   A. sterility is to pelvic inflammatory disease
   B. underdevelopment is to mumps
   C. hyneria is to vaginal disorder
   D. syphilis is to yeast infection

   Ans. A

3. An infertility treatment for women is to embryos from prize sheep can be temporarily implanted into rabbits for easy shipping by air, as:

   A. artificial insemination is to extra-cellular fertilization
   B. artificial insemination is to extra-ovarian conception
   C. fertility drug is to interspecific reimplantation
   D. oocyte donation is to host mothering

   Ans. D

4. What is it called if the nucleus is removed from an ovum and replaced with the nucleus of a cell from an adult donor?

   A. cloning
   B. nuclear transplant
   C. ovarian manipulation technique
   D. ovarian reinplantation

   Ans. A

Development

5. All of the changes that take place during the entire life of an animal from conception to death, is referred to as:

   A. growth
   B. morphogenesis
   C. development
   D. cell differentiation

   Ans. C

6. What is the name of the process by which cells organize themselves, shaping the multicellular animal with its intricate pattern of tissues and organs?

   A. development
   B. embryology
   C. cellular differentiation
   D. morphogensis

   Ans. D

7. What is the name of the process by which cells become specialized?

   A. cell development
   B. cell differentiation
   C. cell morphogenesis
   D. cell division and growth

   Ans. B

8. The majority of a zygote's cytoplasm and organelles originate from:

   A. both ovum and sperm
   B. the sperm
   C. the ovum
   D. the developing zygote which produces its own organelles and cytoplasm following fertilization

   Ans. C

9. The numbers of chromosomes that a zygote possesses:

   A. are contributed equally by the sperm and ovum
   B. are contributed more by the ovum if the zygote is going to be a female and contributed more by the sperm if the zygote is going to be a male
   C. are contributed more by the ovum regardless of the zygote's sex
   D. are the outgrowth of the zygote which develops its own chromosomes regardless of the parents

   Ans. A

10. A series of rapid, mitotic divisions following fertilization are to a number of cells which form a ball of cells, as:

    A. embryonic development is to gastrula
    B cell differentiation is to blastula
    C. cell differentiation is to gastrula
    D. cleavage is to morula

    Ans. D

11. Several hundred cells which form a hollow ball are to a fluid-filled cavity as:

    A. blastula is to archenteron
    B. morula is to coelom
    C. blastula is to blastocoel
    D. morula is to gastrocoel

    Ans. C

12. In the amphibian egg, fertilization causes rearrangement of some of the superficial cytoplasm that contains dark granules. This shift exposes the underlying lighter-colored cytoplasms which is referred to as:

    A. white crescent
    B. gray crescent
    C. homogenous gray granules
    D. gray partitioned cytoplasm

    Ans. B

13. When the first two cells of a frog embryo are separated experimentally, what becomes of these two cells embryologically?

    A. one develops the anterior structures (head) while the other develops the posterior structures (tail)
    B. one develops into a functional embryo, while the other fails to develop
    C. neither of the cells is capable of normal development because they were separated
    D. both cells develop into complete tadpoles

    Ans. D

14. Yolk consists of a mixture of:

    A. proteins, phospholipids, and fats
    B. proteins, glucose, and glycogen
    C. carbohydrates, lipids, and nucleic acids
    D. proteins, DNA, RNA, glucose, and lipids

    Ans. A

Development

15. Vertebrate eggs that have large amounts of yolk concentrated at one end of the cell are to eggs in which the opposite end is more metabolically active as:

    A. vegetal pole is to living pole
    B. nutritive pole is to living pole
    C. vegetal pole is to animal pole
    D. nutritive pole is to embryonic pole

    Ans. C

16. Radial cleavage is to spiral cleavage as:

    A. protostomes are to deuterostomes
    B. deuterostomes are to protostomes
    C. protostomes are to zygostomes
    D. deuterostomes are to zygostomes

    Ans. B

17. In what types of animals would you find the blastula consists of many small cells in the animal hemisphere and fewer, but larger cells in the vegetable hemisphere?

    A. bony fishes and amphibians
    B. insects and starfishes
    C. reptiles and birds
    D. birds and mammals

    Ans. A

18. In what types of animals would you find that cleavage only takes place in a small disc of cytoplasm at the animal pole?

    A. animals and birds
    B. insects and spiders
    C. fishes and amphibians
    D. reptiles and birds

    Ans. D

19. The process by which an embryo becomes three layered (possesses three germ layers) is referred to as:

    A. blastulation
    B. cleavage stage
    C. gastrulation
    D. morphogenesis

    Ans. C

20. Which is the correct sequence of developmental events?

    A. cleavage, zygote, morula, blastula, and gastrula
    B. cleavage, zygote, blastula, morula, and gastrula
    C. cleavage, zygote, blastula, gastrula, and morula
    D. zygote, cleavage, morula, blastula, and gastrula

    Ans. D

21. The embryonic tissue layers are referred to as:

    A. formative layers
    B. germ layers
    C. dermal layers
    D. tissue genesis layers

    Ans. B

22. The primitive gut or developing digestive cavity is to the lining of the developing digestive cavity, as:

    A. archenteron is to mesoderm
    B. gastrula is to mesoderm
    C. archenteron is to endoderm
    D. stomach is to mesoderm

    Ans. C

23. The nervous system and sense organs are to skeletal tissue and muscle, as:

    A. ectoderm is to mesoderm
    B. mesoderm is to endoderm
    C. mesoderm is to ectoderm
    D. ectoderm is to endoderm

    Ans. A

24. When cells at the vegetal pole flatten and begin to invaginate, the process is referred to as:

    A. morphogenesis
    B. blastulation
    C. differentiation
    D. gastrulation

    Ans. D

Development

25. The opening of the archenteron is to the structure the archenteron becomes in deuterostomes, as:

    A. stoma is to nasal openings
    B. blastopore is to anus
    C. gastrostoma is to mouth
    D. blastopore is to mouth

    Ans. B

26. What structure induces or stimulates the development of the nervous system?

    A. notochord
    B. ectoderm
    C. mesoderm
    D. archenteron

    Ans. A

27. Central cells of the neural plate that move downward are to cells of the neural plate that move upward, as:

    A. neural folds are to neural tubes
    B. neural folds are to notochord
    C. neural groove is to neural folds
    D. notochord is to neural tube

    Ans. C

28. Membranes found in terrestrial animals that enclose the entire embryo are to an outgrowth of the developing digestive tract that stores nitrogenous wastes in birds and reptiles, as:

    A. allantois and chorion are to amnion
    B. amnion and chorion are to yolk sac
    C. allantois and chorion are to yolk sac
    D. amnion and chorion are to allantois

    Ans. D

29. The human gestation period averages how many days from the time of conception?

    A. 256 days
    B. 266 days
    C. 280 days
    D. 286 days

    Ans. B

30. Eventually forms the chorion and amnion are to the structure that gives rise to the embryo proper, as:

    A. blastocyst is to trophoblast
    B. trophoblast is to blastocyst
    C. trophoblast is to inner cell mass
    D. inner cell mass is to blastocyst

    Ans. C

31. Implantation occurs in what structure?

    A. endometrium of the uterus
    B. mesoderm of the uterus
    C. endoderm of the uterus
    D. endoderm of the zona pellucida

    Ans. A

32. The organ of exchange between the mother and the embryo is to the embryonic structure from which the organ of exchange between the mother and embryo develops, as:

    A. placenta is to chorion
    B. umbilical cord is to chorion
    C. umbilical cord is to amnion
    D. villi are to amnion

    Ans. A

33. Human chorionic gonadotropin signals what structure within the mother that pregnancy has begun?

    A. uterus
    B. corpus luteum
    C. follicle
    D. anterior pituitary gland

    Ans. B

34. The process of birth is to a long series of involuntary contractions of the uterus, as:

    A. labor is to cervical dilation
    B. labor is to uterine reflex action
    C. parturition is to labor
    D. born is to parturition

    Ans. C

Development

35. The first breath of a neonate is thought to be initiated by the accumulation of:

    A. oxygen
    B. carbon dioxide
    C. nitrogen
    D. methane

    Ans. B

36. What is the name of a photograph taken of the embryo by using ultrasound?

    A. ultragram
    B. sonar imaging
    C. sonar diagnosis
    D. sonogram

    Ans. D

37. Trace the development of an *Amphious* embryo from zygote to gastrula, including drawings of each stage.

38. Compare cleavage in an *Amphioxus* to an amphibian.

39. Why do terrestrial vertebrates develop an amnion?

40. Explain human blastocyst formation and implantation.

Chapter 44

Animal Behavior

1. Some animals have evolved a precise evolutionary adaptation to seasonal changes in climate, availability of food resources and safe nesting sites. This adaptation is referred to as:

   A. operant conditioning
   B. circadian rhythms
   C. homeostasis
   D. migration

   Ans. D

2. The responses of an organism to signals from its environment, including those from other organisms, are referred to as:

   A. adaptations
   B. behavior
   C. experiences
   D. reflexes

   Ans. B

3. Structure, function, and behavior all interact and relate as mechanisms that help define an organism and equip it for survival. All of these characteristics are referred to as:

   A. adaptations
   B. developmental responses and mechanisms
   C. fitness responses and mechanisms
   D. selective agents

   Ans. A

4. The study of behavior in natural environments from the point of view of adaptation, is referred to as:

   A. ecology
   B. elytronology
   C. ethology
   D. entomology

   Ans. C

Animal Behavior

5. The daily rhythmic activities that some organisms exhibit, are referred to as:

   A. cichlid rhythms
   B. circadian rhythms
   C. cistron rhythms
   D. caecilian rhythms

   Ans. B

6. Animals active during the day are to animals busiest during the twilight hours, as:

   A. diurnal is to nocturnal
   B. diurnal is to circadian
   C. photophilic is to nocturnal
   D. diurnal is to crepuscular

   Ans. D

7. Behavior involves all body systems, but it is primarily influenced by two systems, which are:

   A. nervous and circulatory systems
   B. digestive and circulatory systems
   C. nervous and endocrine systems
   D. endocrine and circulatory systems

   Ans. C

8. Instinct is to a stimulus that elicits a fixed action pattern (FAP), as:

   A. behavior is to conditioning
   B. innate is to sign stimulus
   C. imprinting is to learning
   D. behavior is to insight learning

   Ans. B

9. A physiologically meaningful stimulus is to an irrelevant stimulus, as:

   A. unconditioned stimulus is to conditioned stimulus
   B. conditioned stimulus is to learning
   C. learning is to imprinting
   D. habituation is to insight learning

   Ans. A

10. When an animal must do something in order to gain a reward, it is referred to as:

    A. habituation
    B. insight learning
    C. imprinting
    D. operant conditioning

    Ans. D

11. To gain a reward or to avoid punishment is to removal of a stimulus increases the probability that a behavior will occur, as:

    A. negative reinforcement is to positive reinforcement
    B. positive reinforcement is to negative reinforcement
    C. imprint learning is to insight learning
    D. imprint learning is to task related learning

    Ans. B

12. For animals to learn to perform complex tasks like walking or perfecting feeding skills, the process involved is probably:

    A. imprint learning
    B. classical conditioning
    C. operant conditioning
    D. fixed action pattern

    Ans. C

13. The bond, which is usually formed within a few hours of birth or hatching, forms by a type of learning known as:

    A. survival learning
    B. insight learning
    C. habituation
    D. imprinting

    Ans. D

14. An animal learns to ignore a repeated, irrelevant stimulus is to the most complex learning, which is the ability to adapt past experiences that may involve different stimuli to solve a new problem, as:

    A. habituation is to insight learning
    B. survival learning is to response learning
    C. stimulus learning is to cognitive learning
    D. survival learning is to cognitive learning

    Ans. A

15. What organ in migrating birds senses changes in day length, to trigger migratory behavior?

    A. hypothalamus
    B. pituitary
    C. pineal gland
    D. adrenal gland

    Ans. C

16. The direction of travel is important in animal navigation. Honey bees and sea turtles appear to be sensitive to:

    A. light
    B. magnetic fields
    C. temperatures
    D. olfactory signals

    Ans. B

17. Birds appear to navigate by a combination of:

    A. celestial, and geographical and climate cues
    B. learning experiences from older birds that have previously completed migrations
    C. celestial, and olfactory cues
    D. celestial, and magnetic fields and infrared radiation

    Ans. A

18. When animals maximize energy obtained per unit of foraging time, they maximize their reproductive success. The most efficient way for an animal to obtain food, is referred to as:

    A. efficient ingestion
    B. adaptive foraging
    C. selective foraging
    D. optimal foraging

    Ans. D

19. Many factors of the physical environment bring animals together into groups that are not necessarily social. Such a group or gathering of animals is referred to as:

    A. population
    B. aggregation
    C. community
    D. society

    Ans. B

20. The interaction of two or more animals, usually of the same species, is referred to as:

    A. aggregate behavior
    B. population
    C. social behavior
    D. community

    Ans. C

21. An actively cooperating group of individuals belonging to the same species, are referred to as:

    A. society
    B. aggregation
    C. population
    D. pack or flock

    Ans. A

22. Increased competition for food and habitats, and the increased risk of transmitting disease, are costs that are associated with:

    A. populations
    B. social behavior
    C. communities
    D. ecosystems

    Ans. B

23. One animal can influence the behavior of another only if they can exchange mutually recognizable signals, a mechanism referred to as:

    A. social behavior
    B. learning behavior
    C. insight learning
    D. communication

    Ans. D

24. Chemical signals that convey information between members of a species, are referred to as:

    A. hormones
    B. enzymes
    C. pheromones
    D. steroid olfactory stimulator

    Ans. C

Animal Behavior

25. What is the name of the organization within a society that regulates aggressive behavior within the group, due to each individual's arrangement of status?

    A. aggression suppression
    B. dominance hierarchy
    C. imprint behavior
    D. instinctive behavior

    Ans. B

26. What sex hormone can increase aggressiveness among some individuals, and can increase an indiviual's status within a society?

    A. testosterone
    B. estrogen
    C. progesterone
    D. adrenalin

    Ans. A

27. A geographical area that an animal seldom or never leaves is to a portion of the home range that an animal will defend against other individuals of their own species, as:

    A. habitat is to environmental home
    B. habitat is to territory
    C. ecosystem is to territory
    D. home range is to territory

    Ans. D

28. Cooperation between two individuals, the temporary suppression of aggressive behavior, and a system of communication are indicative of behavior that is referred to as:

    A. parental behavior
    B. sexual behavior
    C. imprint behavior
    D. learning behavior

    Ans. B

29. A stable relationship between animals of the opposite sex that ensures cooperative behavior in mating and rearing of the young is referred to as:

    A. mate bonding
    B. parental bonding
    C. pair bonding
    D. family bonding

    Ans. C

30. Specific cues enable courtship rituals to function as mechanisms, referred to as:

    A. reproductive isolating mechanisms
    B. species specific mechanisms
    C. pair bonding mechanisms
    D. parental bonding mechanisms

    Ans. A

31. Females improve their fitness by:

    A. mating with several males
    B. being highly selective regarding mate selection
    C. investing little energy in the reproductive process and more in growth
    D. investing a minimal amount of energy in each reproductive effort, so they can reproduce over an extend period of time

    Ans. B

32. Males improve their fitness by:

    A. being highly selective regarding mate selection
    B. mating with one female, and putting all of their energy into the success of one reproductive effort
    C. allocating most of their energy for growth and less for reproduction, thus they are more competitive
    D. mating with as many females as possible

    Ans. D

33. Their job within the society is reproduction is to their job is to forage, as:

    A. queen is to younger workers
    B. older workers are to younger workers
    C. queen bee is to older workers
    D. queen bee is to soldiers

    Ans. C

34. Physically and behaviorally specialized castes are to a great range and plasticity of potential behavior, as:

    A. termites and fishes are to primates and rodents
    B. birds and fishes are to primates and rodents
    C. bees and wasps are to termites and ants
    D. termites and ants are to bees and wasps

    Ans. A

Animal Behavior

35. One individual appears to act in such a way as to benefit others rather than itself is to the evolutionary effect caused by individuals who favor the survival and reproduction of their relatives, as:

    A. kin selection is to group selection
    B. altruisic behavior is to group selection
    C. group selection is to social selection
    D. altruistic behavior is to kin selection

    Ans. D

36. The evolution of social behavior through natural selection, is referred to as:

    A. ethology
    B. sociobiology
    C. social ecology
    D. evolutionary biology

    Ans. B

37. Why have some species evolved diurnal behavior, while other species have evolved nocturnal behavior?

38. Differentiate between an organized society and an aggregation of organisms.

39. Explain why territoriality has evolved among some species.

40. Explain why courtship rituals have evolved.

Chapter 45

Ecology Populations

1. Environmental deteriorations, hunger, persistent poverty, and health issues are all factors that are associated with:

   A. global warming
   B. increasing human populations
   C. decreasing economy
   D. lack of health care

   Ans. B

2. Countries are categorized according to the level of economic prosperity, and their population growth rate. How many categories are commonly used?

   A. two
   B. three
   C. four
   D. five

   Ans. B

3. A single child born in the United States has the equivalent impact on the environment and resource utilization as how many children born in Niger or Bangladesh?

   A. two
   B. five
   C. twelve
   D. thirty

   Ans. C

4. Ecology is the study of:

   A. nature
   B. pollution produced by humans, the litter indicative of our "throw away" society, and consequent recycling.
   C. wildlife conservation and the varying aspects of forestry
   D. interactions among and between organisms and their physical environment

   Ans. D

5. A group of organisms that belong to the same species that inhabit the same area at the same time, is referred to as:

   A. population
   B. community
   C. ecosystem
   D. adaptive ecological unit

   Ans. A

6. The gaseous envelope that surrounds the Earth is referred to as:

   A. lithosphere
   B. gasosphere
   C. atmosphere
   D. aerosphere

   Ans. C

7. The interaction between all the populations in a specific area or region, is:

   A. ecosystem
   B. ecosphere
   C. biosphere
   D. community

   Ans. D

8. The soil and rock of the Earth's crust, is referred to as:

   A. geosphere
   B. lithosphere
   C. episphere
   D. hypogeansphere

   Ans. B

9. An ecosystem is defined as:

   A. all the communities of living things on Earth
   B. all the interactions between the biosphere and their physical environment
   C. all the interactions among living organisms of a community and their interactions with the physical environment
   D. all the interactions among living organisms of a population and their interactions with the physical environment

   Ans. C

10. The number of individuals of a species per unit of area at a given time, is referred to as:

    A. population density
    B. population carrying capacity
    C. exponential growth
    D. biotic potential

    Ans. A

11. Nesting seabirds, plants that produce inhibitory chemicals that prevent other plants from germinating in the surrounding area, and animals that exhibit territorial behavior, typically exhibit what type of dispersions?

    A. clumped dispersion
    B. random dispersion
    C. nonrandom dispersion
    D. uniform dispersion

    Ans. D

12. Organisms like the flour beetle larvae that occupy an unusually homogeneous environment, exhibit what type of dispersion?

    A. clumped dispersion
    B. random dispersion
    C. nonrandom dispersion
    D. uniform dispersion

    Ans. B

13. The type of dispersion that typically occurs as a result of inefficient seed dispersal or asexual reproduction in plants is referred to as:

    A. clumped dispersion
    B. random dispersion
    C. nonrandom dispersion
    D. uniform dispersion

    Ans. A

14. What is the growth rate (r=b-d) for a population of 10,000 individuals in which there were 2,000 births per year and 1500 deaths per year?

    A. 5 % growth rate
    B. 10% growth rate
    C. 15% growth rate
    D. 20% growth rate

    Ans. A

15. What is the doubling time ($t_d = {}^{0.7}/r$) for a population of 10,000 individuals in which there were 2,000 births per year and 1500 deaths per year, and the doubling time is calculated as 0.7?

    A. 7 years
    B. 10 years
    C. 14 years
    D. 21 years

    Ans. C

16. When individuals leave a population and thus decrease its size, it is referred to as:

    A. migration
    B. population regulatory mechanism
    C. immigration
    D. emigration

    Ans. D

17. What represents the correct equation for the growth rate of a population if r represents population growth rate, e represents emigration, i represents immigration, b represents birth rate, and d represents deathrate?

    A. r = (b-d) + (e-i)
    B. r = (b-d) + (i-e)
    C. r = (d-b) + (e-i)
    D. r = (d-b) + (i-e)

    Ans. B

18. The maximum rate at which a population could increase under ideal conditions is known as its:

    A. population growth rate
    B. population potential
    C. biotic potential
    D. maximum biological growth rate

    Ans. C

19. The maximum rate at which a population could increase under ideal conditions would be the smallest for:

    A. bacteria
    B. mice
    C. rabbits
    D. whales

    Ans. D

20. A population growth curve that has a "J" shape is characteristic of:

A. exponential growth
B. stabilized growth
C. negative growth
D. geometrical growth

Ans. A

21. When a population's growth rate decreases to around zero, and the population approaches the limits of the environment to support a population it is referred to as:

A. the environmental maximum or (EM)
B. the environmental maximum capacity or (EMC)
C. the carrying capacity or (K)
D. the population maximum potential or (PMP)

Ans. C

22. If protozoans (one-celled animals) contained within a beaker reproduced unchecked for an extended period of time, they would eventually run out of food and living space, and poisonous metabolic wastes would accumulate. The population would begin to decrease, due to the collective affects of the factors listed above, which are referred to as:

A. population decreases
B. environmental resistance
C. negative population regulators
D. population growth limiters

Ans. B

23. Predation, disease, and competition are examples of factors that limit population growth, which specifically are referred to as

A. population growth inhibitors
B. independent regulatory agents
C. population suppressors
D. density-dependent factors

Ans. D

24. Blizzards, hurricanes, and fire, are examples of factors that limit population growth, which specifically are referred to as:

A. nonspecific population regulators
B. population influencing agents
C. density-independent factors
D. independent regulatory agents

Ans. C

25. In predator-prey interactions, such as the relationship between the Canada Lynx and the snowshoe hare, their populations show:

    A. similar population fluctuations with a lag time in the response of the predator to the prey
    B. similar population fluctuations with a lag time in the response of the prey to the predator
    C. similar population fluctuations, with the predator responding directly to prey fluctuations
    D. no similar fluctuations, indicating that the two populations are independent on one another's population fluctuations.

26. When members of the same species compete for limited resources, such as food, water, living space, sunlight, and mates, it is referred to as:

    A. interspecific competition
    B. species competitive interaction
    C. population resource competition
    D. intraspecific competition

    Ans. D

27. Organisms that have evolved r selected traits exhibit:

    A. low population growth rates
    B. high population growth rates
    C. fluctuating population growth rates, that are dependent upon the availability of resources
    D. unpredictable population growth rates, because their populations do not respond to the availability of resources

    Ans. B

28. Organisms that have evolved r selected traits exhibit:

    A. large body size, and reproduce once, producing a large number of progeny or offspring
    B. large body size, and reproduce many times, producing a large number of progeny or offspring each time
    C. small body size, and reproduce many times, producing a large number of progeny or offspring
    D. small body size, and reproduce many times, producing a small number of progeny or offspring

    Ans. C

29. Organisms that have evolved K selected traits exhibit:

    A. low population growth rates
    B. random population growth rates that are independent of density influencing factors
    C. high population growth rates that are very dependent upon resources
    D. high population growth rates in which individuals allocate a large amount of energy to reproduce

    Ans. A

30. Organisms that have evolved K selected traits exhibit:

    A. a preference for relatively constant or stable environments, where they are selected for low competitive ability
    B. preference for fluctuating or unstable environments, where they are selected for higher competitive ability
    C. a preference for fluctuating or unstable environments where selective pressure has been relaxed, thus competitive ability is low
    D. a preference for relatively constant or stable environments, where they have a high competitive ability

    Ans. D

31. Presently, the human population has been increasing 1 billion individuals every:

    A. 6 years
    B. 12 years
    C. 24 years
    D. 36 years

    Ans. B

32. Thomas Malthus, a British economist recognized that human populations cannot continue to increase indefinitely. He pointed out that human population growth was capable of:

    A. equaling the rate at which food is produced
    B. increasing faster than was the food supply
    C. avoiding problems associated with growth if famine and war are prevented
    D. increased growth, but economic factors would cause it to eventually attain stability

    Ans. B

33. When the birth rate equals the death rate, this is referred to as:

    A. population stability
    B. population equilibrium
    C. dynamic population equilibrium
    D. zero population growth

    Ans. D

34. Worldwide, the average number of children born to a woman during her lifetime, is 3.2. This number of children is referred to as:

    A. replacement-level fertility
    B. replacement population growth
    C. total fertility rate
    D. population growth fertility

    Ans. C

35. When birth and death rates are high, and population grows at a moderate rate, it is indicative of what demographic stage?

    A. pre-industrial stage
    B. post-industrial stage
    C. industrial stage
    D. transitional stage

    Ans. A

36. The geographic limit of a population's distribution is referred to as its:

    A. dispersion
    B. territory
    C. range
    D. geographic population factors

    Ans. C

37. Explain why some populations exhibit a J-shaped population growth curve, while others exhibit an S-shaped population growth curve.

38. Explain when individuals would be selectively favored for r selection and when K selection would be favored.

39. When would it be advantageous for individuals to be evenly dispersed, and when is clumped dispersion advantageous?

40. Explain how a population can have an increase in its growth rate when the birth rate is declining.

Chapter 46

Communities of Organisms

1. Due to changing selective pressures, a species that cannibalizes its own young for a major part of its food supply will:

   A. probably have to change it's food supply to another species of animals
   B. probably switch from a carnivorous existence to an omnivorous existence
   C. probably cause its own extinction
   D. probably cause its population to decrease to a very low level, until other species increase and the population can increase again

   Ans. C

2. Before the introduction of non-native species into Lake Victoria, the ciclids helped maintain a:

   A. proper balance between other species of fish
   B. proper balance between other species of ciclids
   C. properly balanced water chemistry
   D. proper balance between producers, consumers, and decomposers

   Ans. D

3. The ecological balance of Lake Victoria was primarily caused by:

   A. introduction of numerous, non-native species
   B. introduction of the Nile perch
   C. over-fishing by humans
   D. careless use of herbicides, pesticides, and commercial fertilizers

   Ans. B

4. As ciclid populations declined in Lake Victoria, dense algalblooms began to die, causing:

   A. a loss of habitat for most invertebrates
   B. a decrease in the amount of food available for other fish and invertebrate populations
   C. their decomposition to use up the available dissolved oxygen in deep water
   D. more food to be available for invertebrates, and selecting other selected species of fish

   Ans. C

Communities of Organisms

5. A community refers to:

   A. an association of organisms of different species living and interacting together
   B. an association of members of the same species living and interacting together
   C. an association between several species of organisms and their physical environment
   D. an association between producers and consumers

   Ans. A

6. Communities vary in size, but they:

   A. have precise boundaries and are usually isolated
   B. are well defined, but show signs of overlap
   C. show signs of overlap, but interactions are very limited
   D. are rarely isolated, thus there is considerable overlap

   Ans. D

7. An ecosystem refers to:

   A. a community and its non-living environment
   B. an interacting population and its non-living environment
   C. individuals of a species interacting with their environment

   Ans. A

8. Organisms that are able to synthesize organic molecules from inorganic molecules are called:

   A. heterotrophs
   B. decomposers
   C. detritivors
   D. autotrophs

   Ans. D

9. Organisms that are not capable of synthesizing their own food, and are dependent upon other organisms for their nutritional needs are referred to as

   A. heterotrophs
   B. autotrophs
   C. grannivors
   D. isotrophs

   Ans. A

10. Plant is to herbivore as:

    A. autotroph is to producer
    B. producer is to consumer
    C. producer is to detritivor
    D. consumer is to decomposer

    Ans. B

11. Deer is to wolf as:

    A. autotroph is to heterotroph
    B. carnivore is to herbivore
    C. primary consumer is to secondary consumer
    D. secondary consumer is to omnivore

    Ans. C

12. Vulture is to roadkill, as:

    A. earth is to earthworm
    B. detritivor is to detritus
    C. primary consumer is to secondary consumer
    D. detritus is to detritus feeder

    Ans. B

13. Bear is to salmon and berries, as:

    A. primary consumer is to secondary consumer and producer
    B. heterotroph is to producer and consumer
    C. carnivore is to heterotroph and autotroph
    D. omnivore is to consumer and producer

    Ans. D

14. Saprophytes are to nutritional products released for energy, as:

    A. bacteria and fungi are to decomposition products
    B. algae and protozoans are to inorganic molecules
    C. carbon dioxide and mineral salts are to bacteria and fungi
    D. termites and symbiotic bacteria are to proteins and nucleic acids

    Ans. A

Communities of Organisms

15. What do autotrophs produce for all living organisms?

    A. nutrients and amine (NH^2) groups
    B. carbon dioxide and organic compounds
    C. food and oxygen
    D. potassium, nitrogen, and phosphorous

    Ans. C

16. What prevents dead organisms and waste products from accumulating indefinitely?

    A. they are taken up by the producers and their roots
    B. earthworms are the primary agents that dispose of these materials
    C. insects consume most of the materials in terrestrial environments and crayfish and bottom feeding fish assume this role in aquatic environments
    D. the decomposers

    Ans. D

17. When a community is carefully analyzed, and the roles of the numerous organisms are evaluated, it becomes evident that:

    A. producers can live independently of all other organisms
    B. decomposers can live independently of all other organisms
    C. if light is available, photosynthesizers can exist independently
    D. no organism exists independent of other living organisms

    Ans. D

18. The three main types of interactions which occur among species in a community are:

    A. photosynthesis, reproduction, and herbivory
    B. competition, predation, and symbiosis
    C. autotrophic reactions, consumer reactions, and decomposition
    D. photosynthesis, decomposition, and reproduction

    Ans. B

19. The interdependent evolution between two interacting species, such as a predator exerting selective pressure on its prey, while the prey evolves better strategies to escape, and thus applies pressure on the predators, is referred to as:

    A. coevolution
    B. counter selection
    C. predator-prey selective reaction
    D. counter evolution

    Ans. A

20. Two of the most common predator strategies that have evolved for capturing prey are:

    A. sharp teeth and acute eyesight
    B. speed and strength
    C. pursuit and ambush
    D. large size and acute sense of smell

    Ans. C

21. Prey that have evolved warning coloration to advertize their unpalatability have colors that are:

    A. classified as camouflage
    B. are usually green
    C. are usually brown or tan
    D. are bright, contrasting

    Ans. D

22. When a harmless or edible species has evolved a resemblance to a harmful or inedible species, the strategy is referred to as:

    A. Mullerian mimicry
    B. Batesian mimicry
    C. Aggressive mimicry
    D. Automimicry

    Ans. B

23. When two different species, all of which are poisonous, harmful, or distasteful, have evolved resemblance to one another the strategy is referred to as:

    A. Mullerian mimicry
    B. Batesian mimicry
    C. Mertesian mimicry
    D. Aggressive mimicry

    Ans. A

24. When two or more species interact in a relationship, or association that may benefit two species, or a relationship in which two species may be unaffected, or one species is harmed by the relationship, this is collectively called:

    A. endosymbiosis
    B. symbiosis
    C. ectosymbiosis
    D. interspecific evolved interaction

    Ans. B

Communities of Organisms

25. The relationship that has evolved between reef-building coral animals and algae called Zooanthellae is referred to as:

    A. commensalism
    B. parasitism
    C. mutualism
    D. epiphytic symbiosis

    Ans. C

26. The relationship that evolved between small plants that are called epiphytes, and the larger tropical trees in which they live attached to the bark is referred to as:

    A. mutualism
    B. endosymbiosis
    C. parasitism
    D. commensalism

    Ans. D

27. The distinctive role a species has within the structure and function of an ecosystem, is referred to as:

    A. its habitat
    B. its adaptation
    C. its niche
    D. biological adaptive habitat

    Ans. C

28. When environmental stress is low, species diversity is

    A. high
    B. intermediate
    C. low
    D. can not be predicted

    Ans. A

29. When comparing two distinctive communities which adjoin one another, species diversity would most probably be highest in what region?

    A. in the center of each community
    B. in the transitional zone
    C. in the area furthest from the region where the two communities adjoin
    D. it would be the same throughout each community, regardless of the region

    Ans. B

30. The process of community development over time, which involves species in one stage being replaced by different species is referred to as:

    A. climax community
    B. community interaction gradient
    C. succession
    D. community time-scale interactive replacement profile

    Ans. C

31. It has been found that no two species can indefinitely occupy the same niche in the same community because eventually a process will occur that will replace one of the species, due to interspecific competition. This process of replacing one species is referred to as:

    A. niche replacement
    B. coexistence replacement interaction
    C. minimizing overlap co-efficient
    D. competitive exclusion

    Ans. D

32. When an environmental variable tends to restrict the ecological niche of a particular organism it is referred to as:

    A. an adaptive environmental factor
    B. an adaptive tolerance
    C. limiting factor
    D. ecological restricter

    Ans. C

33. The change in species composition over time in a habitat that has not previously been inhabited by organisms, such as a bare rock that becomes inhabited by lichens, is referred to as:

    A. primary succession
    B. pioneer succession
    C. environmental succession
    D. abiotic-biotic succession

    Ans. A

Communities of Organisms

34. When abandoned farmland begins to change in species composition, and a new community begins to develop in a predictable manner it is referred to as:

    A. postcultivation succession
    B. secondary succession
    C. pioneer succession
    D. herbaceous plant succession

    Ans. B

35. The ecological niche of an organism may be far broader potentially than it is in actuality, or an organism is usually capable of utilizing much more of its environment's resources than it actually does. This potential niche is referred to as its:

    A. fundamental niche
    B. specific niche
    C. realized niche
    D. competitive niche

    Ans. A

36. When one species benefits from a relationship with another species, and the second species is adversely affected, it is referred to as:

    A. commensalism
    B. pathogen interaction
    C. parasitism
    D. predation

    Ans. C

37. Differentiate between a population, community, and ecosystem.

38. Differentiate between the following:
    A. consumer and decomposer
    B. decomposer and producer
    C. producer and consumer

39. Differentiate between the following:
    A. Aposomatic coloration and Batesian mimicry
    B. Batesian mimicry and Mullerian mimicry
    C. Crypsis or camouflage prey and warning coloration

40. Why is the realized niche of an organism generally narrower than its fundamental niche?

Chapter 47

Ecosystems

1. When the increasing concentration of pesticides in the tissues of organisms higher in the food chain occurs, it is referred to as:

   A. pesticide multiplication
   B. biological magnification
   C. biological pesticide tolerance
   D. exponential pesticide growth

   Ans. B

2. DDT was especially detrimental and hazardous to birds because:

   A. they died of pesticide toxicity from low concentrations of the pesticide
   B. they developed blindness from very low concentrations of the pesticides
   C. the pesticide caused their eggs to be extremely thin and fragile, and usually resulted in breakage during incubation
   D. of their sensitivity to the pesticide. It causes their metabolism to be altered, which results in continuous loss of weight and eventually terminates in death

   Ans. C

3. Some pesticides, particularly chlorinated hydrocarbons are extremely stable and may take years to break down into less toxic forms. The fact that they take such a long time to break down is referred to as:

   A. persistence
   B. negative degradation
   C. environmental toxic accumulation
   D. long term pesticide residue

   Ans. A

4. Chlorinated hydrocarbons, such as DDT that cannot be metabolized or excreted are usually stored in the organism's

   A. liver
   B. nerve tissue
   C. bone marrow
   D. fatty tissues

   Ans. D

Ecosystems

5. When an ecologist studies an ecosystem they study:

   A. the specific habitat of an organism
   B. the interactions of a species with their environment
   C. the inactions of a community with their abiotic environment
   D. the adaptations of a population and how they evolved in relation to the abiotic environment

   Ans. C

6. The ecosphere is comprised of:

   A. biosphere and its interactions with the hydrosphere, lithosphere, and atmosphere
   B. biosphere and its interaction with the hydrosphere, geosphere, and chemosphere
   C. biosphere and its interactions with the hydrosphere, thermosphere, and heliosphere
   D. biosphere and its interactions with the thermosphere, solarsphere, and hydrosphere

   Ans. A

7. The four different biogeochemical cycles of matter that are representative of all biogeochemical cycles are:

   A. water, carbon, oxygen, and nitrogen
   B. carbon, hydrogen, oxygen, and nitrogen
   C. carbon, potassium, phosphorous, and oxygen
   D. carbon, nitrogen, phosphorous, and water

   Ans. D

8. Proteins, carbohydrates, lipids, and nucleic acids all contain carbon. Carbon is present:

   A. only in living organisms as a result of cellular respiration
   B. in the atmosphere, water, and rocks
   C. as carbon dioxide involved as a reactant in photosynthesis
   D. as a product of the greenhouse effect

   Ans. B

9. The overall equation for photosynthesis is:

   A. $12\ CO_2 + 12\ H_2O$ >>light>> $2\ C_6H_{12}O_6 + 6O_2 + 6\ H_2O$
   B. $6\ CO_2 + 18\ H_2O$ >>light>> $C_6\ H_{12}\ O_6 + 6\ O_2 + 12\ H_2O$
   C. $6\ CO_2 + 12\ H_2O$ >>light>> $C_6H_{12}O_4 + 6\ O_2 + 12\ H_2O$
   D. $5\ CO_2 + 10\ H_2O$ >>light>> $C_5H_{10}O5 + 5\ O_2 + 10\ H_2O$

   Ans. C

10. The overall equation for aerobic respiration is:

    A. $C_6H_{12}O_6 + 6\ O_2 + 12\ H_2O$-----> $6\ CO_2 + 18\ H_2O$ + energy for biological work
    B. $C_6H_{12}O_6 + 6\ O_2 + 6\ H_2O$ ---> $6\ CO_2 + 6\ H_2O$ + energy for biological work
    C. $C_5H_{10}O_5 + 5\ O_2 + 5\ H_2O$ ---->$5\ CO_2 + 10\ H_2O$ + energy for biological work
    D. $C_6H_{12}O_6 + 6\ O_2 + 6\ H_2O$----> $6\ CO_2 + 12\ H_2O$ + energy for work

    Ans. D

11. Carbon dioxide fixation is to carbon dioxide release as:

    A. cellular respiration is to photosynthesis
    B. biological cycle is to abiotic cycle
    C. photosynthesis is to cellular respiration
    D. producer is to consumer

    Ans. C

12. Conversion of gaseous compound is to organism that converts compound to biological compound, as:

    A. nitrogen fixation is to bacteria
    B. nitrogen fixation is to plants
    C. hydrogen fixation is to plants
    D. phosphorous fixation is to bacteria

    Ans. A

13. Loss of water vapor is to living organism, as:

    A. photosynthesis is to plant
    B. transpiration is to plant
    C. respiration is to animal
    D. evaporation is to estuary

    Ans. B

14. Evaporation and precipitation are to the area of land being drained by runoff, as

    A. transpiration is to groundwater
    B. transpiration is to estuary
    C. global warming is to oceans
    D. hydrologic cycle is to watershed

    Ans. D

Ecosystems

15. Energy from food passes from one organism to the next in a sequence is to a range of choices of food on the part of each organism involved, as:

    A. producer is to omnivor
    B. producer is to consumer
    C. food chain is to food web
    D. trophic level is to food pyramid

    Ans. C

16. The first trophic level is to the third trophic level as:

    A. primary consumer is to carnivore
    B. producer is to decomposer
    C. producer is to omnivor
    D. producer is to secondary consumer

    Ans. D

17. Each step in a food web is to the sum or total number of organisms at each nutritional level in a constructed figure, as:

    A. food chain is to biomass
    B. trophic level is to an ecological pyramid
    C. food chain is to primary and secondary consumers
    D. trophic level is to community structure

    Ans. B

18. Coal, oil, and natural gases are collectively referred to as:

    A. utility generating fuels
    B. energy producing fuels
    C. fossil fuels
    D. combustible non-renewable resources

    Ans. C

19. As levels of atmospheric carbon dioxide slowly and steadily increase, this may result in changes in climate, referred to as:

    A. global warming
    B. global atmospheric alteration
    C. biosphere warming
    D. ecosphere warming

    Ans. A

20. Nitrogen fixation involves the conversion of gaseous nitrogen (N_2) to:

    A. ammonia (NH_3)
    B. nitrite (NO_2)
    C. nitrate (NO_3)
    D. amine (NH_2)

    Ans. A

21. Nitrification is the process in which ammonia is converted to:

    A. amino acids
    B. nitrites (NO_2)
    C. nitrates (NO_3)
    D. amines (NH_2)

    Ans. C

22. Assimilation is the process in the nitrogen cycle in which plant roots absorb either:

    A. nitrate (NO_3) or nitrite (NO_2)
    B. nitrate (NO_3) or ammonia (NH_3)
    C. nitrite (NO_2) or ammonia (NH_3)
    D. urea or uric acid

    Ans. B

23. Commercial fertilizer in itself is not bad, but its overuse on the land can cause water pollution. Nitrate fertilizer washes from fields into rivers and lakes where it:

    A. has a toxic effect on fish
    B. absorbs dissolved oxygen (O)
    C. stimulates an excessive growth of bacteria, which decompose and increase the levels of carbon dioxide
    D. stimulates an excessive growth of algae, which die, and their decomposition uses up dissolved oxygen

    Ans. D

24. Manure of sea birds is to large amounts of natural fertilizer as:

    A. bird dropping is to ammonia (NH_3) and nitrite (NO_2)
    B. guano is to ammonia (NH_3) and nitrate (NO_3)
    C. guano is to phosphate and nitrate
    D. FeCa is to iron and calcium

    Ans. C

Ecosystems

25. Energy flow refers to the passage of energy in a one-way direction through an ecosystem. The correct sequence of energy flow is:

    A. sunlight, photosynthesis, energy stored in organic molecules, cellular respiration, energy release
    B. sunlight, photosynthesis, energy stored in inorganic molecules, energy and organic compounds released
    C. sunlight, energy stored photosynthesis, energy released
    D. organic molecules synthesized by bacteria, absorbed by plants, sunlight, photosynthesis, energy released

    Ans. A

26. The rate at which energy is captured and stored in plant tissues during photosynthesis is to the amount of energy that remains in plant tissues after cellular respiration has occurred, as:

    A. primary production is to tertiary production
    B. gross primary production is to net primary production
    C. net primary production is to gross primary production
    D. primary production is to secondary production

    Ans. B

27. Climate refers to the:

    A. the average temperature for a region over a period of years
    B. the average precipitation for a region over a ten year period of time
    C. the average temperature and precipitation for a region over an extended period of time
    D. the average weather conditions that occur over a period of years

    Ans. D

28. The sun's energy is a product of:

    A. the burning of helium
    B. the burning of helium and carbon which forms an unstable reaction
    C. a massive nuclear fusion reaction
    D. a nuclear entrophy reaction

    Ans. C

29. The sun's energy which is emitted into space is in the form of electromagnetic radiation, especially:

    A. cosmic rays, x-rays, and light
    B. light, infrared, and ultraviolet radiation
    C. light and ultraviolet radiation, and radio waves
    D. light radiation, heat, and magnetic waves

    Ans. B

30. The Earth's roughly spherical shape and tilted angle of its axis produces a great deal of variation in the exposure of the Earth's surface, which results in:

    A. variation in temperature
    B. sunlight hitting the equator obliquely
    C. sunlight hitting the arctic pole vertically in the summer and indirectly in the winter
    D. sunlight causing the greatest tides to occur at the two poles

    Ans. A

31. Seasons are determined by:

    A. the inclination of the sun as the Earth rotates
    B. the prevailing winds blowing over the oceans
    C. the inclination of the Earth's axis as it rotates the sun
    D. by the effects of gyros

    Ans. C

32. A rain shadow refers to an area in which there is:

    A. high precipitation
    B. periodic precipitation
    C. seasonal periods of drought
    D. low precipitation

    Ans. D

33. Rain shadows occur on which side of a mountain range?

    A. windward side
    B. north facing slopes
    C. leeward side
    D. south facing slopes

    Ans. C

Ecosystems

34. Precipitation in mountains occurs primarily on:

    A. south facing slopes
    B. windward slopes
    C. north facing slopes
    D. leeward slopes

    Ans. B

35. The direction of energy flow through an ecosystem is described as:

    A. polydirectional
    B. bimodal
    C. progressive
    D. linear

    Ans. D

36. Ecological pyramids express reduction in energy at:

    A. lower trophic levels
    B. primary trophic levels
    C. decomposer trophic levels
    D. higher trophic levels

    Ans. D

37. Why is a food web considered more realistic than a food chain?

38. Diagram the hydrologic cycle.

39. Diagram the three different types of food pyramids.

40. Why is the cycling of matter essential to all ecosystems?

Chapter 48

Major Ecosystems of the World

1. Predominant rangeland vegetation is to root system as:

   A. forbs are to taproots
   B. shrubs are to taproots
   C. grasses are to fibrous roots
   D. grasses are to adventitious roots

   Ans. C

2. Maximum number of animals that rangeland can sustain is to the results that occur when the maximum capacity is exceeded, as:

   A. carrying capacity is to overgrazing
   B. standard grazing capacity is to excess grazing capacity
   C. rangeland determined grazing is to macrograzing limit
   D. K value is to negative K value

   Ans. A

3. Rapid deterioration of rangeland due to poor management practices, often results in:

   A. infertile rangeland
   B. lithification
   C. erosional sedimentation
   D. desertification

   Ans. D

4. In the African Sahel, the desert has expanded as a result of:

   A. extended drought conditions
   B. human overpopulation
   C. lack of genetic diversity among the rangeland plants
   D. extensive fires attributed to human slash and burn techniques

   Ans. B

Major Ecosystems of the World

5. A large relatively distinct ecosystem that is characterized by similar climate, soil, plants, and animals is to the northern most distinct ecosystem that some plants can tolerate, as:

   A. ecotone is to artic ice cap
   B. ecosphere is to tundra
   C. biome is to tundra
   D. ecosphere is to taiga

   Ans. C

6. Permafrost refers to soil that is characterized by:

   A. permanently frozen throughout the year
   B. surface soil that thaws during the summer, but has deep layers of permanently frozen ground
   C. containing geologically old soil due to its location
   D. soil that thaws quickly and erodes extremely fast

   Ans. B

7. Characteristic permafrost plants are to permafrost produced landscape as:

   A. large deep rooted plants are to large deep lakes
   B. small plants and lichens are to fast flowing streams and lakes
   C. shallow rooted conifers are to shallow lakes and bogs
   D. small plants and lichens are to sluggish streams and bogs

   Ans. D

8. Animals found on the tundra are to animals absent from the tundra, as:

   A. lemmings, voles, and ptarmigan are to frogs, salamanders, and snakes
   B. lemmings, squirrels, and toads are to turtles, lizards, and bats
   C. snowy owls, crows, and elk are to weasels, artic fox, and raccoons
   D. musk-oxen, caribou, and small frogs are to large frogs, snakes, and hares

   Ans. A

9. North American and Eurasian boreal forest is to drought-resistant adaptations, as:

   A. deciduous trees are to thick leaves and cuticles
   B. taiga is to needle-like leaves with minimal surface area
   C. evergreen is to thin leaves with short petioles
   D. taiga is to deep root system that can withstand long periods of drought

   Ans. B

10. In boreal forests roots can not absorb water when:

    A. the soil is waterlogged because it presents hypotonic condition
    B. the soil is saturated due to acidity
    C. the ground is frozen
    D. the air temperature is less than the soil temperature

    Ans. C

11. The boreal forest is not well suited to agriculture because:

    A. it has a short growing season and is mineral poor
    B. it is too wet and has a high clay content
    C. it is too sandy and is prone towards drought
    D. it is too cold for agricultural crops and is too acidic

    Ans. A

12. The climate of the North American continent, especially the West, is dominated by:

    A. dry northerly winds
    B. considerable precipitation throughout the entire area
    C. sparse precipitation throughout
    D. rain shadows

    Ans. D

13. The coniferous temperate rain forest occurs in North America on the:

    A. Northeast coast of North America
    B. Southeast coast of North America
    C. Northwest coast of North America
    D. South central coast of North America

    Ans. C

14. Dominant vegetation in the North American temperate rain forest is to nonparasitic smaller plants associated with larger plants, as:

    A. lodgepole pine and subalpine fir is to herbaceous dicots
    B. Douglas fir and Sitka spruce is to epiphytes
    C. Engelmann spruce and ponderosa pine is to dwarf mistletoe

    Ans. B

Major Ecosystems of the World

15. The logging practice preferred by foresters or the logging industry is:

    A. selective cutting
    B. uneven age cutting
    C. partial selective cutting
    D. clear cutting

    Ans. D

16. When the logging industry replants an area that has been harvested, they replant the area as:

    A. a forest community
    B. a polyculture that consists of multiple woody species
    C. a monoculture
    D. they frequently allow the area to undergo natural selection

    Ans. C

17. Temperate deciduous forests are adapted to what kind of conditions?

    A. seasonality
    B. mild conditions with little seasonal variation
    C. mild conditions with moderate precipitation
    D. mild conditions with abundant precipitation

    Ans. A

18. Temperate deciduous forests are dominated by:

    A. evergreen conifers
    B. deciduous conifers
    C. broad-leaved trees
    D. nearly equal numbers of conifers and broad-leaved trees

    Ans. C

19. In Europe, Asia, and America, deciduous forests have been converted into agricultural use. The results have been:

    A. very similar for all three areas, because cultivation has had little effect
    B. quite different, because in Europe and Asia, traditional agricultural methods were employed without a substantial loss of fertility; but in America, they abandoned wise soil conservation practices and allowed erosion
    C. quite different, because in Asia they traditionally utilize prudent agricultural practices that focus on soil conservation, while in Europe and America, farmers have depended upon artificial fertilizers, herbicides, and pesticides which have destroyed the community
    D. very similar for all these regions, because these soils are poor to begin with, thus the entire community has been disrupted and destroyed beyond recovery

    Ans. B

20. Temperate grasslands and deciduous forests have characteristic temperatures and annual precipitation which are:

    A. grasslands have hot summers and cold winters, and both have 20 to 25 inches of precipitation annually
    B. grasslands have cool summers and cold winters, while forests have hot summers and cold winters, and both have 50 to 70 inches of precipitation annually
    C. grasslands have mild summers and winters with 19 to 30 inches of precipitation, while forests have mild summers and winters with 30 to 50 inches of precipitation annually
    D. grasslands have hot summers and cold winters with 10 to 30 inches of precipitation, while forests have hot summers and cold winters with 30 to 50 inches of precipitation annually

    Ans. D

21. Most of the plants that are adapted to temperate grasslands are classified as:

    A. annuals
    B. biannuals
    C. perennials
    D. hydrophytes

    Ans. C

Major Ecosystems of the World

22. Most of the plants that are adapted to temperate grasslands are:

    A. adapted to wet weather
    B. adapted to fire
    C. adapted to moderate conditions, as opposed to extremes such as drought, hot temperatures, cold weather and wind
    D. moderately adapted to uplands, but primarily adapted to lowlands like river bottoms and along streams

    Ans. B

23. The height of prairie grasses, such as tallgrass prairie and shortgrass prairie, is primarily due to what selective agent?

    A. amount of moisture
    B. amount of sunlight
    C. amount of grazing pressure applied by large herbivorous animals
    D. amount of temperature variatiation

    Ans. A

24. The temperate habitats that are classified as chaparral are similar to climates found

    A. in the Sahara desert
    B. in the Amazon basin
    C. around the Mediterranean Sea
    D. throughout the temperate grasslands

    Ans. C

25. Chaparrals habitats characteristically have climates that are:

    A. warm, dry winters and cool, wet summers
    B. hot, dry winters and hot, dry summers
    C. mild, dry winters and mild, wet summers
    D. mild, wet winters and very dry summers

    Ans. D

26. Desert temperatures are to desert soil as:

    A. little temperature fluctuations are to high organic content and low mineral content in the soil
    B. extremely hot and cold temperatures are to low organic and high mineral content of soils
    C. extremely hot and cold temperatures are to high organic and mineral content of the soil
    D. extremely hot days and moderate nights are to low organic and mineral content of the soil

    Ans. D

27. Desert leaf adaptations are to chemical substances that inhibit establishment of competing plants as:

    A. leaves that are reduced or absent, are to chemopathy
    B. leaves that are broad, flat, and near the ground, are to inhabitase
    C. leaves that are reduced or absent are to allelopathy
    D. leaves that conserve water are to toxophytopathy

    Ans. C

28. Size of desert animals is to daily activity schedule, as:

    A. generally small is to nocturnal activity
    B. generally large is to nocturnal activity
    C. generally large is to diurnal
    D. generally small is to diurnal

    Ans. A

29. Savanna typically has a climate which is characterized by:

    A. moderate rainfall and few droughts
    B. low or seasonal rainfall with prolonged dry periods
    C. abundant rainfall throughout the year, but soil that does not retain moisture
    D. abundant rainfall throughout the year, but soil is very compact thus there is extreme soil erosion

30. Tropical rain forest plants are to the distinct vegetative stories as:

    A. deciduous flowering plants are to two distinct vegetative stories
    B. evergreen flowering plants are to two distinct vegetative stories
    C. deciduous flowering and nonflowering plants are to three distinct vegetative stories
    D. evergreen flowering plants are to three or more distinct stories of vegetation

    Ans. D

31. Tropical rain forest epiphytes are to abundant tropical rain forest animals, as:

    A. woody vines or lianas are to insects, amphibians, and reptiles
    B. conifers are to sloths and monkeys
    C. orchids and bromeliads are to insects, amphibians, and reptiles
    D. byrophytes and lichens are to insects, spiders, and amphibians

    Ans. C

Major Ecosystems of the World

32. The life zones that might be encountered in Colorado, beginning at the lowest elevation and continuing to the top of a 14,000 foot peak, occur in the following sequence:

    A. deciduous forest, coniferous forest, alpine tundra, permanent ice and snow
    B. grasslands, coniferous forest, deciduous forest, and alpine tundra
    C. deciduous forest, grassland, coniferous forest, and permanent ice and snow
    D. deciduous forest, coniferous forest, grassland, and alpine tundra

    Ans. A

33. Degrees north latitude are to increase in altitude, as:

    A. permafrost and intense solar radiation are to lack of permafrost and less intense solar radiation
    B. great extremes of day length and less intense solar radiation are to absence of great extremes of day length and intense solar radiation
    C. greater exposure to ultraviolet radiation and permafrost are to less exposure to ultraviolet radiation and permafrost at high elevations
    D. less extremes of day length and greater exposure to ultraviolet radiation

    Ans. B

34. Plankton are to nekton, as:

    A. strong swimmers are to bottom dwelling organisms
    B. relatively feeble swimmers are to organisms fixed to the bottom
    C. microscopic, relatively feeble swimmers are to larger stronger swimmers
    D. microscopic, relatively feeble swimmers are to organisms adapted to bottom feeding

    Ans. C

35. Small streams that are the sources of a river, are usually shallow, cold, swiftly-flowing and therefore highly oxygenated, are referred to as:

    A. lotic brooks
    B. tributary creeks
    C. lentic rivers
    D. headwater streams

    Ans. D

36. Organisms found in fast flowing water ecosystems may have evolved adaptations such as:

    A. flattened bodies and suckers for attachment
    B. larger swim bladders for buoyancy and streamlined bodies
    C. vertically flattened on a dorsal-ventral axis and large appendages
    D. vertically flattened bodies or bodies flattened on a dorsal-ventral axis, and suckers for attachment

    Ans. A

37. Why are estuaries extremely valuable economically and biologically?

38. How could desertification be reversed in the African Sahel?

39. Why are most animals in tropical rain forests adapted to the trees?

40. What effect does the construction of a dam have upon organisms that inhabit a stream with a fast current?

Chapter 49

Environmental Problems

1. Dusky seaside sparrow, Abingdon tortoise, Dodo, and Giant Auk are to San Joaquin Kit Fox, Black Rhinoceros, Pitcher plant, and Whooping Crane, as:

   A. species at risk are to species that are extinct
   B. species that are no longer at risk are to species at risk
   C. species that are extinct are to species that are at risk
   D. species that are no longer at risk are to species that are extinct

   Ans. C

2. The Earth's biological diversity is currently:

   A. decreasing slightly
   B. decreasing at an alarming rate
   C. remains in equilibrium
   D. increasing slightly

   Ans. B

3. As many as one-fourth of the higher plant families may be:

   A. increasing their numbers by the beginning of the 21st century
   B. increasing their numbers by the end of the 21st century
   C. extinct by the beginning of the 21st century
   D. extinct by the end of the 21st century

   Ans. D

4. Directly attributable to human activities are to occurring in a tremendously compressed period of time, as:

   A. extinction is to rate of extinction occurring within a few years as opposed to hundred of years
   B. rate of species increase is to rate of species increase is occurring with a few decades due to changes in habitat
   C. rate of species equilibrium is to rate at which species equilibrium occurs, is within decades as opposed to hundreds of years
   D. mass extinction is to rate of mass extinction that occurs within a few decades as opposed to millions of years

   Ans. D

Environmental Problems

5. The extinction of plants has severe consequences because:

   A. plants are the bases or foundations of food webs
   B. plants require longer periods of time to produce than animals
   C. plants are more susceptible to environmental changes than animals
   D. plants do not have the extensive ranges that animals do

   Ans. A

6. A species' numbers are so severely reduced that it is in danger of becoming extinct are to a species' population is low, but extinction is less eminent, as:

   A. threatened is to at risk
   B. at risk is to threatened
   C. at risk is to endangered
   D. endangered is to threatened

   Ans. D

7. Any species whose numbers have been severely reduced represent a decline in biological diversity because:

   A. they occupy less habitat
   B. their genetic variability is severely diminished
   C. their rate of mutation increases due to small numbers
   D. they become more interdependent upon other species

   Ans. B

8. The primary cause of species endangerment and extinction is:

   A. too much collecting and harvest
   B. disease and parasitism
   C. destruction of natural habitat
   D. hunting

   Ans. C

9. The most probable cause for the decline of large stands of forest trees and the biological death of many freshwater lakes is:

   A. industrial pollution
   B. acid precipitation
   C. agricultural pollution
   D. sewage

   Ans. B

10. The balance among native organisms living in an area may be upset by:

    A. the introduction of a foreign or exotic species
    B. overprotecting the species and allowing too many individuals to reproduce
    C. over protecting the species and not letting natural selective agents act upon them
    D. overprotecting the species and not allowing diseases and parasites to selectively act upon them

    Ans. A

11. Predator and pest control are to commercial hunting, as:

    A. prairie dogs and Carolina parakeets are to mountain lions and grizzly bears
    B. wolves and tigers are to pocket gophers and black-footed ferrets
    C. black-footed ferrets and snow leopards are to cheetah and mountain lions
    D. pocket gophers and mountain lions are to cheetahs and tigers

    Ans. D

12. Dagger handles and aphrodisiacs are to furs and gall bladders, as:

    A. elephant tusks are to black-footed ferrets and tigers
    B. rhinoceros horns are to snow leopards and bears
    C. bear teeth and claws are to prairie dogs and wolves
    D. elephants tusks are to prairie dogs and wolves

    Ans. B

13. Establishing parks and reserves for preserving biological diversity in the wild is referred to as:

    A. natural species preservation
    B. aided species conservation
    C. in situ conservation
    D. ex situ conservation

    Ans. C

14. Conserving biological diversity in human-controlled settings, is referred to as:

    A. zoological parks conservation
    B. breeding captive species conservation
    C. in situ conservation
    D. ex situ conservation

    Ans. D

Environmental Problems

15. An area that is set aside that may serve recreational needs is to areas set aside for recreational use, but is also used for grazing and farming operations, as:

    A. national parks are to national forests
    B. national forests are to wildlife refuges
    C. wildlife refuges are to national parks
    D. national forests are to national parks

    Ans. A

16. Sperm collected from a suitable male is collected and used to impregnate a female in some distant location is to fertilized eggs are surgically implanted into a female of a related species which is less rare, as:

    A. host mothering is to embryo transfer
    B. artificial insemination is to embryo transfer
    C. host mothering is to artificial insemination
    D. embryo transfer is to artificial species preservation

    Ans. B

17. The best method to prevent species from becoming endangered is:

    A. decrease hunting and collecting
    B. place a few individuals in zoos and botanical gardens
    C. increase reclamation projects
    D. protect natural habitat

    Ans. D

18. Permanent destruction of all tree cover is to the consequences of destroying tree cover, as:

    A. clearcutting is to increasing habitat
    B. slash and burn is to increasing habitat
    C. deforestation is to decreased soil fertility and increased erosion
    D. deforestation is to increased soil nutrient cycling

    Ans. C

19. When trees release substantial amounts of moisture into the air, the process is referred to as:

    A. evaporation
    B. transpiration
    C. photosynthesis
    D. hydrologic cycling

    Ans. B

20. Regional and global climate changes are induced by:

A. deforestation
B. mining and oil exploration
C. decreasing soil fertility
D. increasing soil fertility

Ans. A

21. Subsistence agriculture, commercial logging, and cattle ranching, are the main causes for:

A. increased nutrient cycling
B. decreasing national debt in lesser developed countries
C. deforestation
D. stabilization of economic growth of agricultural and wood products

Ans. C

22. South and Central America, Central Africa, and Southeast Asia, are to 50 percent, as:

A. area of greatest poverty is to the percentage of the human populations in those areas
B. area of greatest loss of habitat is to the percentage of the human population in those regions
C. area of greatest loss of species diversity is to the percentage of the human populations in these regions
D. area where most of the tropical rain forests are located are to the area where most of the Earth's species are found

Ans. D

23. Many species of North American migratory birds are declining in numbers due to:

A. hunting and collecting in both North America and South America
B. loss of habitat in the tropics
C. widespread disease
D. heavy pollution in the tropics

Ans. B

24. By destroying tropical rainforests, we may be reducing or eliminating nature's ability to restore its species through:

A. community natural selection
B. population adaptive evolution
C. adaptive radiation
D. ecosystem evolution

Ans. C

Environmental Problems

25. Dry tropical rain forests are being destroyed primarily for:

    A. use as fuel
    B. agricultural use
    C. commercial logging
    D. conversion into villages

    Ans. A

26. Tropical forests destroyed to create rangeland, progress through ecological succession are to a product produced from many tropical forests that is extremely energy inefficient, as:

    A. prairie is to ethyl alcohol
    B. prairie is to isoproyl alcohol
    C. desert is to herbaceous plants
    D. scrub savannah is to charcoal

    Ans. D

27. Releases of carbon dioxide (CO_2) and N_2O into the atmosphere triggers the production of:

    A. O_2
    B. O_3
    C. CH_4
    D. NH_3

    Ans. B

28. The main contributing factor to the release of CFC's is:

    A. leaking refrigerators and air conditioners
    B. decomposition in landfills
    C. emissions from feedlots
    D. burning of tropical forests

    Ans. A

29. Gases that retain heat in the atmosphere are to the subsequent consequences of retaining these gases, as:

    A. organic gases are to global warming
    B. heat radiate gases are to global warming
    C. greenhouse gases are to greenhouse effect
    D. atmospheric gases are to global warming

    Ans. C

30. Most heat absorbed by stratospheric carbon dioxide:

    A. escapes into space
    B. re-radiates back to Earth
    C. is absorbed by carbon dioxide in the troposphere
    D. combines with methane to form the troposphere

    Ans. B

31. Global warming could cause:

    A. worldwide drought
    B. worldwide increases in precipitation
    C. decrease in food production in the oceans
    D. sea level to rise and cause severe flooding

    Ans. C

32. Ecosystems that will be most severely affected by global warming will be:

    A. tropical forests, tundra, and coastal wetlands
    B. tropical forests, boreal forests, and polar seas
    C. coral reefs, mountain ecosystems, and boreal forests
    D. deserts, temperate forests, and coral reefs

    Ans. C

33. Increasing carbon dioxide in the atmosphere and the resulting increase in temperature will probably not harm human health directly. Human health will be indirectly affected by:

    A. malaria-infested mosquitos and encephalitis-infected flies
    B. rodents that carry plague and respiratory disease caused by increased levels of ozone
    C. botulism and pneumonia
    D. diphtheria and malnutrition due to flooding of agricultural lands

    Ans. A

34. Cause of ozone depletion is to result of ozone depletion, as:

    A. carbon dioxide is to increase in infrared radiation
    B. methane is to increase in hydrocarbons
    C. CFC's are to increase in ultraviolet radiation
    D. carbon dioxide is to re-radiation of heat

    Ans. C

Environmental Problems

35. The highest depletion of stratospheric ozone has been over:

    A. Southeast Asia
    B. Equator
    C. Artic
    D. Antarctica

    Ans. D

36. The base of the food web for the Southern Ocean is:

    A. fish
    B. phytoplankton
    C. whales and seals
    D. crustaceans

    Ans. B

37. Why should we make every effort to protect biological diversity?

38. Why does habitat destruction contribute to declining biological diversity?

39. Explain the environmental costs of deforestation.

40. Explain how the greenhouse effect works, and its consequences.